38

MICROBIOLOGICAL RISK
ASSESSMENT SERIES

T0073452

Listeria monocytogenes in ready-to-eat (RTE) foods: attribution, characterization and monitoring

MEETING REPORT

Food and Agriculture Organization of the United Nations
World Health Organization

Rome, 2022

Required citation:

FAO and WHO. 2022. Listeria monocytogenes *in ready-to-eat (RTE) foods: attribution, characterization and monitoring* – Meeting report. Microbiological Risk Assessment Series No. 38. Rome. https://doi.org/10.4060/cc2400en

ISSN 1726-5274 [Print]
ISSN 1728-0605 [Online]
FAO ISBN 978-92-5-136983-8
WHO ISBN 978-92-4-003496-9 (electronic version)
WHO ISBN 978-92-4-003497-6 (print version)
© FAO and WHO, 2022

Cover picture © Dennis Kunkel Microscopy, Inc

Contents

TABLES

FIGURES

Acknowledgements

The Food and Agriculture Organization of the United Nations (FAO) and the World Health Organization (WHO) would like to express their appreciation to all those who contributed to the preparation of this report through the provision of their time and expertise and other relevant information before, during and after the meeting. Special appreciation is extended to all the members of the expert group for their dedication to this project, in particular to Dr Catherine Donnelly for her leadership in chairing the meeting, and to Dr Jeffrey Farber for his excellent support in preparing the final document. All contributors are listed in the following pages.

Appreciation is also extended to all those who responded to the calls for data that were issued by FAO and WHO and brought to our attention data in official documentation or not readily available in the mainstream literature.

Contributors

EXPERTS

Ana Allende, Senior Researcher, CEBAS-CSIC (Spanish National Research Council), Spain

Sukhadeo B. Barbuddhe, Director, ICAR-National Research Centre on Meat, India

Brecht Devleesschauwer, Senior Epidemiologist, Sciensano, Department of Epidemiology and Public Health, Belgium

Qingli Dong, Professor, University of Shanghai for Science and Technology, China

Catherine Donnelly, Professor of Nutrition and Food Science, the University of Vermont, United States of America

Jeffrey Farber, Adjunct Professor, University of Guelph, Department of Food Science, Canada; Director JM Farber Global Food Safety, Canada

Lisbeth Truelstrup Hansen, Professor, National Food Institute (DTU Food), Technical University of Denmark, Denmark

Alejandra Latorre, Associate Professor, College of Veterinary Sciences, Universidad de Concepción, Chile

Alexandre Leclercq, Deputy Head of the French Reference Centre and WHO Collaborating Centre *Listeria,* Institut Pasteur, France

Kudakwashe Magwedere, Directorate of Veterinary Public Health, Department of Agriculture, Land Reform and Rural Development, South Africa

Deon Mahoney, Principal, Deon Mahoney Consulting, Australia

Tom Ross, Professor in Food Microbiology (Centre for Food Safety and Innovation), Director, ARC Training Centre for Innovative Horticultural Products, Tasmanian Institute of Agriculture (TIA), Australia

Elliot Ryser, Professor, Department of Food Science and Human Nutrition, Michigan State University, United States of America

Marcel Zwietering, Professor in Food Microbiology, Wageningen University, Netherlands

RESOURCE PERSONS

Martin Wiedmann, Gellert Family Professor in Food Safety, Department of Food Science, Cornell University, United States of America

Anne Brisabois, Research Director, Laboratory for Food Safety, ANSES, France

Dorothy-Jean McCoubrey, Director, Dorothy-Jean & Associates Ltd, New Zealand

Jose Emilio Esteban, Chief Scientist, United States Department of Agriculture, United States of America

Verna Carolissen, Food Standards Officer, Joint FAO/WHO Food Standards Programme, Italy

Sarah Cahill, Senior Food Standards Officer, Joint FAO/WHO Food Standards Programme, Italy

Lingping Zhang, Food Standards Officer, Joint FAO/WHO Food Standards Programme, Italy

Goro Maruno, Food Standards Officer, Joint FAO/WHO Food Standards Programme, Italy

Secretariat

Haruka Igarashi, Department of Nutrition and Food Safety, World Health Organization, Switzerland

Christine Kopko, Food Systems and Food Safety Division, Food and Agriculture Organization of the United Nations, Italy

Jeffrey LeJeune, Food Systems and Food Safety Division, Food and Agriculture Organization of the United Nations, Italy

Kang Zhou, Food Systems and Food Safety Division, Food and Agriculture Organization of the United Nations, Italy

Declaration of interests

All participants completed a Declaration of Interests form in advance of the meeting.

One of the experts declared interest in the topic under consideration. The investment interest of Martin Wiedmann is over the FAO/WHO threshold. His declared interests may be considered potentially significant. Anne Brisabois and Dorothy-Jean McCoubrey were unable to attend the meeting in the final days because of prior commitments. Therefore, while they were invited to participate in the meeting, they participated as technical resource people and were excluded from the decision-making process regarding the final recommendations. The rest of the declared interests reported were not considered by FAO and WHO to present any conflict in light of the objectives of the meeting.

All the declarations, together with any updates, were made known and available to all the participants at the beginning of the meeting.

All the experts participated in their individual capacities and not as representatives of their countries, governments or employers.

Abbreviations and acronyms

AMR	antimicrobial resistance
AMR	Region of the Americas (WHO classification)
ANSES	French Agency for Food, Environmental and Occupational Health and Safety
CAC	Codex Alimentarius Commission
CA	competent authority(ies)
CCFH	Codex Committee on Food Hygiene
CDC	Centers for Disease Control and Prevention (United States of America)
CFU	colony-forming unit
DALY	disability-adjusted life year
ECDC	European Centre for Disease Prevention and Control
EEA	European Economic Area
EFSA	European Food Safety Authority
EU	European Union
FAO	Food and Agricultural Organization of the United Nations
FBD	foodborne diseases
FBO	food business operator(s)
FDA	Food and Drug Administration (United States of America)
FERG	Foodborne Disease Burden Epidemiology Reference Group
FSIS	United States Department of Agriculture (USDA), Food Safety and Inspection Services
FVH	fruits, vegetables or herbs
GI	gastro-intestinal

HACCP	hazard analysis critical control point (system)
ID_{50}	dose of an infectious organism required to produce infection in 50 percent of the experimental subjects or exposed population
JEMRA	Joint FAO/WHO Expert Meetings on Microbiological Risk Assessment
LD_{50}	The amount of an infectious organism or toxic agent that is lethal for 50 percent of the exposed population within a certain time
LLO	Listeriolysin O
LMIC	low- and middle-income countries
MOA	Ministry of Agriculture
MPD	maximum population density
MPN	most probable number
MRA	Microbiological Risk Assessment Series
NRC	National Reference Center
NRL	National Reference Laboratory
PHA	Public Health Agency
QMRA	Quantitative microbiological risk assessment
RTE	ready-to-eat
USFDA	United States Food and Drug Administration
WHO	World Health Organization
WGS	whole genome sequencing
YLDs	Years lived with disability
YLLs	Years of life lost

Executive summary

A virtual meeting of the Joint FAO/WHO Expert Meeting on Microbiological Risk Assessment (JEMRA) of *Listeria monocytogenes* (hereinafter referred to as "*L. monocytogenes*") in ready-to-eat (RTE) foods: attribution, characterization and monitoring was held from 20 October to 6 November 2020. The purpose of the meeting was to review recent data on *L. monocytogenes* and determine the need to modify, update, or develop new risk assessment models and tools for this pathogen. A public call for data and experts was issued to support this work. In addition, background documents on the various aspects related to the meeting were prepared ahead of time for consultation by the experts. Prepared documents included the following: 1) assessment of past JEMRA documentation; "Risk assessment of *Listeria monocytogenes* in ready to eat foods: Interpretative summary (MRA4)" (FAO and WHO, 2004a) and "Risk assessment of *Listeria monocytogenes* in ready to eat foods: Technical report" (MRA5) (FAO and WHO, 2004b); 2) a review of current national *L. monocytogenes* surveillance programmes; 3) a review of current microbiological and laboratory methods for *L. monocytogenes*; and 4) an update on the virulence markers for *L. monocytogenes*. The meeting participants reviewed the prepared summary documents and other information on outbreaks and disease attribution, virulence, population risk factors, advances in laboratory methods and surveillance. The aforementioned risk assessment documents (MRA4, MRA5) (FAO and WHO, 2004a, 2004b) covered a cross-section of RTE foods (pasteurized milk, ice cream, cold smoked fish and fermented meats) linked to invasive listeriosis. Since the publication of these documents, outbreaks of listeriosis continue to occur across the globe associated with previously reported foods, but also with many previously unreported food vehicles, including fresh and minimally processed fruits and vegetables (e.g. frozen vegetables). The expert group concluded that it would be wise to be more inclusive in future risk assessments and that a full farm-to-fork risk assessment be considered.

L. monocytogenes can infect anyone; however, it continues to disproportionally affect certain highly susceptible populations. The expert group recommended that future risk assessments should review groupings of susceptible groups, based on physiological risks and other socio-economic factors.

New information has emerged on *L. monocytogenes* strain variants, which differ in their virulence and environmental tolerance. Based on a panel of specific genes, the expert group proposed a virulence ranking of *L. monocytogenes* relevant to invasive listeriosis. The expert group concluded that the development and implementation of effective surveillance systems are critical in addressing the control of *L. monocytogenes*. The use of approved standardized laboratory methods that culture and isolate strains should be the foundation so that human, food and environmental isolates can be further characterized and inventoried.

In conclusion, the expert group identified several critical gaps in the current FAO/WHO risk assessment model and collectively agreed that updating the model would be valuable for informing risk analysis strategies, including in low- and middle-income countries (LMICs). The experts prepared short examples from literature (Annex 1) to demonstrate and highlight several key principles that should be considered in the risk assessment for *L. monocytogenes*.

Introduction

In response to the request from the Codex for scientific advice, FAO and WHO have published several risk assessments on *L. monocytogenes* in foods since 1999 (FAO and WHO, 1999). The work started with fish products (FAO, 1999), and then focused on ready-to-eat (RTE) foods (FAO and WHO, 2000). Risk assessments, previously developed at the national level, were adapted or expanded to address concerns in RTE foods at an international level. To support this work, the 2004 FAO/WHO risk assessment on *L. monocytogenes* (FAO and WHO, 2004a, 2004b) provided scientific insight into the risk characterization of *L. monocytogenes* contamination in food and the seriousness of listeriosis for susceptible populations. The technical report was limited to a cross-section of RTE foods known to cause human listeriosis such as pasteurized milk, ice cream, cold-smoked fish and fermented meats, and the likelihood of these products as vehicles for human foodborne listeriosis.

Since the publication of the 2004 risk assessment, outbreaks of illness and resultant deaths due to *L. monocytogenes* continue to occur across the globe, with the largest one having occurred in South Africa between 2017 and 2018 linked to the consumption of RTE meat products (polony) where a total of 937 laboratory-confirmed cases and 193 deaths were reported. Both polony and environmental samples were found to contain *L. monocytogenes* 4b isolates belonging to ST6, which, together with the isolates from the patients, belonged to the same core-genome multilocus sequence typing cluster with no more than four allelic differences (Thomas *et al.*, 2020). This South African outbreak is the largest and most deadly outbreak of listeriosis recorded globally to-date. Other notable outbreaks have been linked to vehicles not previously identified in the 2004 WHO/FAO risk assessment, including lettuce, packaged salads, cantaloupe or rockmelons, stone fruit, caramel apples, celery, mung bean sprouts and frozen vegetables. An updated risk assessment that considers an examination of produce vehicles is needed.

Continued effort is also needed to summarize and critically evaluate the most recent information on *L. monocytogenes* in RTE foods. New data to improve and further inform the 2004 Risk Assessment is available for nearly every factor considered previously, including new quantitative data on *L. monocytogenes* contamination of foods. An outbreak linked to ice cream suggests that listeriosis may occur after widespread distribution of products that are unable to support growth, but are contaminated at low levels, are consumed by highly susceptible persons. However, it should be noted that a detailed examination of the outbreak strongly suggested that all known exposures related to this outbreak were likely due to the consumption of milkshakes prepared from the original ice cream product, in which case growth of the pathogen could have occurred between preparation and consumption. Additional information is needed to better understand the dose response for highly susceptible subpopulations as well as the relative potential of a single strain of *L. monocytogenes* to evolve to cause severe disease in humans based on virulence gene content and sequences. An updated virulence ranking of *L. monocytogenes* obtained by determining and analysing subtyping data could potentially improve risk assessments. Demographic shifts and changes in the food system, coupled with evidence of listeriosis as a problem in low-income countries, also support an important need for an updated risk assessment.

To facilitate this work, an FAO/WHO expert meeting was held by virtual means from 20 October to 6 November 2020 to review and discuss the available data and background documents, and to assess the need to modify and update risk assessment models/tools. This report focuses on the deliberations and conclusions of the expert meeting.

New research findings and data representing different food commodities and geographical regions will provide opportunities to validate the current risk assessment models for *L. monocytogenes,* assess their application to other food commodities, and potentially develop new risk management approaches to control *L. monocytogenes.*

2

The global burden of foodborne disease associated with *L. monocytogenes*

Foodborne diseases (FBD) represent a constant threat to public health and a significant impediment to socioeconomic development worldwide. However, the priority placed upon food safety, and on specific FBD, varies between countries. A major obstacle to adequately addressing food safety concerns in some jurisdictions is the lack of accurate data on the full extent and burden of FBD. In recent decades, the disability-adjusted life year (DALY) has emerged as the key metric to quantify the population health impact of diseases and risk factors. DALYs integrate the impacts of morbidity (years lived with disability) and mortality (years of life lost), to quantify the healthy life years lost compared to an ideal situation where the world is free from disease. For FBDs in particular, which often are associated with a multitude of health effects, DALYs provide a significant added value over more simple metrics such as incidence and mortality. Since the 2000s, an increasing number of countries and institutions have used DALYs to measure the impact of FBDs at the regional, national or global level. In this section, we summarize and discuss the current evidence on the disease burden of listeriosis.

2.1 WHO ESTIMATES OF THE GLOBAL BURDEN OF LISTERIOSIS

2.1.1 WHO Foodborne Disease Burden Epidemiology Reference Group

In 2006, WHO launched an initiative to estimate the global burden of FBD. This initiative was carried forward by the Foodborne Disease Burden Epidemiology

Reference Group (FERG), an expert group comprising more than 100 experts from different regions of the world. FERG quantified the global and regional burden of 31 foodborne hazards, including 11 diarrhoeal disease agents, seven invasive disease agents, ten helminths, and three chemicals and toxins. Baseline epidemiological data were translated into DALYs following a hazard-based approach and an incidence perspective, which assigns the disease burden of acute and chronic health outcomes to the initial incident event. Data gaps were addressed using statistical imputation models, and the proportions of cases by routes of exposure were generated through structured expert elicitation.

In 2015, the FERG activities resulted in the publication of the first-ever estimates of the global and regional burden of FBD (Havelaar *et al.*, 2015). Using 2010 as the reference year, FERG estimated that the 31 foodborne hazards caused 600 million illnesses, resulting in 420 000 deaths and 33 million DALYs, demonstrating that the global burden of FBD is of the same order of magnitude as major infectious diseases such as HIV/AIDS, malaria and tuberculosis (Havelaar *et al.*, 2015). The burden is also comparable to that related to diet, unsafe water sources (e.g. surface water, unprotected spring water, hand dug wells close to pit toilets, sewer pipes, garbage dumping pits, livestock, etc.) and air pollution. Some hazards were found to be important causes of FBD in all regions of the world, whereas others were highly focal, resulting in a high local burden. Despite the data gaps and limitations linked to these initial estimates, it is evident that the global burden of FBD is considerable, and while it affects individuals of all ages, children under the age of five and persons living in low-income regions are the most affected. Public health officials at the regional, national and international levels can use these estimates to support evidence-based improvements in food safety to improve population health.

2.1.2 Global burden of listeriosis

The FERG estimates of the global and regional burden of listeriosis are presented by Kirk *et al.* (2015), while the methodology for quantifying the global and regional burden of listeriosis was presented in more detail by Maertens de Noordhout *et al.* (2014) (Figure 1). The global and regional incidence of listeriosis was estimated using a multilevel meta-analysis of incidence data obtained through a systematic review of national surveillance data and peer-reviewed and grey literature. The disease model distinguished between perinatal and non-perinatal cases, and considered stillbirths, death, septicemia, central nervous system (CNS) infection, and neurological sequelae following CNS infection. All listeriosis cases were assumed to be foodborne.

The systematic review identified incidence data from 45 out of 194 WHO Member States, mainly high-income countries in the European, American and Western Pacific regions. No incidence data were identified from the African, Eastern Mediterranean and South-East Asian regions.

Incidence data used for estimating
the global burden of listeriosis
▨ Incidence data available
  Incidence data not available
\\\ Excluded from data searches

  Not applicable

FIGURE 1. Geographical coverage of incidence data used for estimating the global burden of listeriosis
Source: Adapted from Maertens de Noordhout *et al.,* 2014.

Overall, listeriosis was estimated to cause 14 169 illnesses in 2010, albeit with a large 95 percent uncertainty interval (UI) ranging from 6 112 to 91 175 cases – reflecting the absence of data from a large part of the world. These illnesses were further estimated to result in 3 175 deaths (95 percent UI 1 339–20 428) and 118 340 DALYs (49 634–754 680). Mainly driven by its low incidence, listeriosis was within the 10 FBDs with the lowest global disease burden. However, it also ranked within the 10 most important FBDs at the patient level, with each case being associated with ~7.5 DALYs. The years of life lost comprised 98 percent of the DALY estimates, reflecting the high case-fatality rate of listeriosis and, in particular, the association with perinatal death and stillbirths.

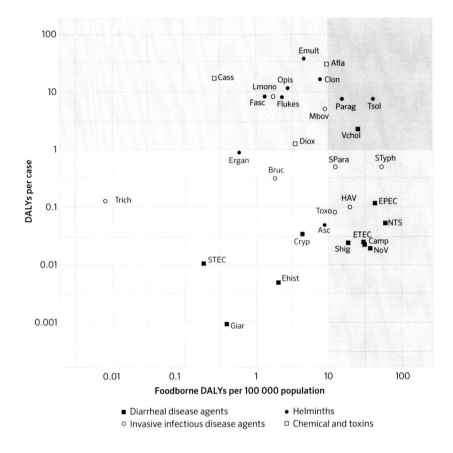

FIGURE 2. Scatterplot of the global burden of foodborne disease per 100 000 population and per case

Source: Adapted with permission from Havelaar *et al.*, 2015.

The grey-shaded areas indicate arbitrary cut-offs between high (H) or low (L) population burden (> or ≤ 10 DALYs per 100 000 population) and high or low individual burden (> or ≤ 1 DALY per case). Abbreviations: NoV: Norovirus; Camp: *Campylobacter* spp.; EPEC: Enteropathogenic *Escherichia coli*; ETEC: Enterotoxigenic *E. coli*; STEC: Shiga toxin-producing *E. coli*; NTS: non-typhoidal *Salmonella enterica*; Shig: *Shigella* spp.; Vchol; *Vibrio cholerae* Ehist: *Entamoeba histolytica*; Cryp: *Cryptosporidium* spp.; Giar: *Giardia* spp.; HAV: Hepatitis A virus; Bruc: *Brucella* spp.; Lmono: *L. monocytogenes*; Mbov: *Mycobacterium bovis*; SPara:

Salmonella Paratyphi A; STyph: *Salmonella* Typhi; Toxo: *Toxoplasma gondii*; Egran: *Echinococcus granulosus*; Emult: *E. multilocularis*; Tsol: *Taenia solium*; Asc: *Ascaris* spp.; Trich: *Trichinella* spp.; Clon: *Clonorchis sinensis*; Fasc: *Fasciola* spp.; Flukes: Intestinal flukes; Opis: *Opisthorchis* spp.; Parag: *Paragonimus* spp.; Diox: Dioxins; Afla: Aflatoxin.

2.2 ESTIMATES OF THE NATIONAL BURDEN OF LISTERIOSIS

The expert group identified national studies on the burden of listeriosis using PubMed, Google Scholar and additional inputs from group members (Table 1). The expert group identified 19 unique publications and reports. The annual updates of the burden of infectious and foodborne disease reports issued by the Dutch National Institute for Public Health and the Environment applied the disease model proposed by Havelaar *et al.* (2012) and were therefore not included as separate entries. The identified studies have been performed in a limited number of high-income countries, in particular in Netherlands (5), United States of America (4), Canada (3), Belgium (2), Germany (1), Greece (1), New Zealand (1), the United Kingdom of Great Britain and Northern Ireland (1), and at the European level (1). Most studies were top-down burden assessments that started from public health surveillance data, while three studies were bottom-up risk assessments that started from exposure linked to specific food sources.

The disease models showed relative similarity across studies. All focused on invasive listeriosis, and most made a distinction between perinatal and non-perinatal (or acquired) listeriosis. The most commonly included non-fatal health conditions were sepsis, meningitis and neurological complications following meningitis. Pneumonia and gastro-enteritis were included in three studies (Kemmeren *et al.*, 2006; Lake *et al.*, 2010, Haagsma *et al.*, 2009).

The estimated DALYs per case ranged from 1.4 to 9.1, with a median of 4.1. The DALYs per case were highly influenced by the relative number of perinatal vs non-perinatal cases and the considered life expectancy table. All studies found a high contribution of the years-of-life lost (YLL) to the DALY estimate, with years lived with disability (YLD) contributing only a minor fraction, or sometimes not even considered.

TABLE 1. National studies on the disease burden of listeriosis

Reference	Setting	Included symptoms and sequelae	DALY/1 000 000	DALY/case
(Kemmeren et al., 2006)	Netherlands	**Perinatal:** death in perinatal period, long-term sequelae **Acquired:** sepsis, meningitis, gastroenteritis, pneumonia, death, long-term sequelae	24.4	5.7
(Haagsma et al., 2009)	Netherlands	**Perinatal:** sepsis, pneumonia, meningitis, long-term sequelae **Acquired:** gastroenteritis, septicemia, meningitis, long-term sequelae	19.1	4.0
(Newsome et al., 2009)	United States of America, associated with whole milk	Not available	N/A	N/A
(Lake et al., 2010)	New Zealand	**Perinatal:** spontaneous abortion/stillbirth, septicemia, meningitis, and pneumonia in surviving neonates, and subsequent long-term neurological sequelae or death of neonates **Acquired:** sepsis, meningitis, gastroenteritis, pneumonia, long-term neurological sequelae, and death	65.7	9.1
(Ruzante et al., 2010)	Canada, associated with ready to eat foods	Not available	1.8	N/A
(Gkogka et al., 2011)	Greece	Not applicable (YLD were not considered)	4.1	4.6
(Chen et al., 2013)	United States of America, associated with soft ripened cheese	**Perinatal:** death in perinatal period, long-term sequelae **Acquired:** sepsis, meningitis, gastroenteritis, pneumonia, death, long-term sequelae	0.1	5.5
(Havelaar et al., 2012)	Netherlands	**Perinatal:** death, meningitis, neurological disorders following meningitis **Acquired:** death, meningitis, neurological disorders following meningitis, acquired listeriosis	6.9	1.4
(Hoffmann et al., 2012); (Batz et al., 2012); (Batz et al., 2014)	United States of America	**Perinatal:** death, hospitalized, chronic **Acquired:** death, no medical care, physician visit, hospitalized	30.6 (QALY losses)	5.9

(cont.)

Reference	Setting	Included symptoms and sequelae	DALY/1 000 000	DALY/case
(Kwong et al., 2012)	Ontario, Canada	*Not available*	3.0	N/A
(Werber et al., 2013)	Germany	*Not applicable (only mortality considered)*	51.5	2.1
(Mangen et al., 2015)	Netherlands	**Perinatal:** death, meningitis, sepsis, neurological sequelae **Acquired:** death, meningitis, neurological sequelae	12.2	2.3
(Scallan et al., 2015)	United States of America	**Perinatal:** death, meningitis, bacteraemia, neurological disorders **Acquired:** death, meningitis, bacteraemia	29.5	5.5
(Thomas et al., 2015)	Canada, 2008 outbreak	**Acquired:** meningitis, bacteraemia, focal infection	4.3	2.5
(van Lier et al., 2016)	Netherlands	**Perinatal:** death, meningitis, neurological disorders following meningitis **Acquired:** death, meningitis, neurological disorders following meningitis, acquired listeriosis	9.6	2.2
(FSA and FFS, 2017)	United Kingdom	**Acquired:** flu-like illness, meningitis, septicaemia	11.4 (QALY losses)	4.0
(Maertens de Noordhout et al., 2017)	Belgium	**Perinatal:** permanent disability following meningitis **Acquired:** symptomatic uncomplicated, symptomatic complicated, permanent disability following meningitis	42.7	7.5
(Cassini et al., 2018)	Europe	**Perinatal:** permanent disability following meningitis **Acquired:** symptomatic uncomplicated, symptomatic complicated, permanent disability following meningitis	22.9	4.1
(Jacquinet et al., 2018)	Belgium	**Perinatal:** permanent disability following meningitis **Acquired:** symptomatic uncomplicated, symptomatic complicated, permanent disability following meningitis	39.7	3.5

2.3 DISCUSSION OF AVAILABLE DISEASE BURDEN ESTIMATES

The FERG study provided the first estimates of the global and regional disease burden of listeriosis. Compared with other foodborne hazards considered, the global burden of listeriosis is moderate; however, the impact at the patient level is considerable, in particular due to the high case-fatality rate. So far, only a limited number of countries have performed national studies on the burden of listeriosis but yield conclusions that are in line with those of the FERG study.

2.3.1 Listeriosis disease model

All available studies on the disease burden of listeriosis considered the central nervous system symptoms and sequelae, while only a few also considered pneumonia and gastro-enteritis.

Pneumonias and pleural infections caused by *L. monocytogenes* are observed in older patients, especially males, with immunosuppression and underlying pleural/pulmonary disease (Morgand *et al.*, 2018). Reflecting the high host vulnerability and the same range of mortality/morbidity than other typical forms of invasive listeriosis, *L. monocytogenes*-associated respiratory infections are rare and appear to be under-reported, as are other atypical *L. monocytogenes* infections.

After the ingestion of food containing high numbers of *L. monocytogenes*, gastroenteritis can appear mainly in immunocompetent individuals after 24 hours and may persist between 1 and 3 days (until one week) (Ooi and Lorber, 2005). Typical manifestations of gastro-intestinal (GI) listeriosis include fever, acute watery diarrhea, nausea, headache, arthralgia and myalgia. Bacteremia rates during GI listeriosis seem to be low (around 2.5 percent) but are not well documented. Several outbreaks of foodborne gastroenteritis due to *L. monocytogenes* have been reported.

Surveillance for gastroenteritis due to *L. monocytogenes* is difficult because the organism is not looked for routinely in cases of gastroenteritis, and there are no gold standard methods for its detection in stool samples. However, given the consistent finding of a predominance of the YLL component in the DALY estimate, including those studies that did consider pneumonia and gastroenteritis, inclusion of these symptoms in the global burden of disease estimates is not considered to have a noticeable impact on the overall results.

2.3.2   Availability of data on the global occurrence of listeriosis

The global and national burden studies reveal the scarcity of listeriosis incidence data across the world. Indeed, incidence data and burden estimates are available only for a limited number of mainly high-income countries. A complete lack of incidence data was identified for the African (except for South Africa), Eastern Mediterranean and South-East Asian regions. In addition, for Latin America, according to the publicly available data on official reports of human cases of listeriosis, information is only available for Chile and Uruguay, where the notification of listeriosis cases is mandatory. This information, however, may not be updated regularly by the local health authorities. For the remaining countries, there is no information about the incidence of listeriosis, with the exception of sporadic cases reported in the literature. This lack of information regarding the actual burden of listeriosis in Latin America is likely attributed to a lack of both specific surveillance and standard reporting of listeriosis (e.g. passive or indirect surveillance through data obtained from mandatory notification of foodborne outbreaks; mandatory notification of foodborne outbreaks in general), rather than an actual low occurrence of listeriosis in the region.

The present review of reported outbreaks occurring between 2005 and 2020 painted a similar picture of data scarcity. Invasive listeriosis is a more severe form of disease which affects certain high-risk groups of the population. Non-Invasive listeriosis usually only results in mild GI illness; however, invasive listeriosis is characterized by meningitis or bacteremia. Infection during pregnancy may result in foetal loss through miscarriage, stillbirth, neonatal meningitis or bacteremia. Laboratory confirmation of invasive infection with symptoms includes the isolation of *L. monocytogenes* from a normally sterile site (e.g. blood, cerebral spinal fluid, joint, pleural or pericardial fluid) or in the case of miscarriage or stillbirth, the isolation of *L. monocytogenes* from placental or foetal tissue (including amniotic fluid and meconium). Only outbreaks of invasive listeriosis that identified a strong connection to a food source were captured as part of this review. Out of 127 identified outbreaks, the majority were reported from the European (69) and American (49) Regions. Nine (9) outbreaks were reported from the Western Pacific Region (one outbreak spanned two WHO regions, the American and Western Pacific Region [United States of America and Australia]), and one (1) from the African Region (South Africa). A total of 3 628 cases of invasive listeriosis occurred in these outbreaks, of which at least 606 (17 percent) were reported as maternofoetal and 230 (6 percent) as occurring in immunocompromised individuals. In total, there were 554 deaths, with a case-fatality rate of 15 percent; at least 27 of the deaths (5 percent) were perinatal. Please see Annex 2 for a detailed analysis of the data from all the listeriosis outbreak investigations.

2.4 CONCLUSIONS

The WHO FERG estimated the global burden of listeriosis in 2010 based on the incidence data representing 48 percent of the world population. Some individual countries have estimated their own burden of listeriosis, but these studies were limited to high-income regions.

The incorporation of new data on the incidence of invasive listeriosis in LMICs (in particular from the African, Eastern Mediterranean and South-East Asian regions), through a systematic review of peer-reviewed studies and national surveillance, would make these estimates more globally representative and more precise.

The relevance of gastroenteritis on the listeriosis burden of disease is not fully understood; however, currently it is considered to be minimal as compared to the bacteremia, meningitis and/or foetal-maternal manifestations observed with invasive listeriosis.

3

Source attribution associated with *L. monocytogenes*

3.1 SOURCE ATTRIBUTION

To guide interventions in the food chain, it is relevant to estimate sources of listeriosis. Source attribution has generally large uncertainties, but especially for listeriosis, source attribution is difficult due to the long incubation time of the disease. The first stage of attribution is the determination of the various major transmission pathways (food, environment, human-to-human, animal, and travel related). Then specifically for RTE food, an attribution can be made for various food groups such as meat, dairy, fish and shellfish as well as fruits and vegetables. Food groups can then again be further subdivided. Several methods of attribution can be used, all having specific advantages and disadvantages (Pires *et al.*, 2009; Mughini-Gras *et al.*, 2019).

The first separation between methods for source attribution is the choice between a top-down or a bottom-up approach (Mughini-Gras *et al.*, 2019). Top-down methods relate cases back to the source, making use of, for example, epidemiological data (outbreak analysis, case control studies) or microbiological data. A bottom-up analysis is based on a risk assessment, where prevalence, concentration and consumption data (exposure assessment) are combined with a dose-response assessment to estimate cases related to one or various food product groups. Additionally, expert elicitations can be used as a further source of information (EFSA, 2014).

TABLE 2. Description of various types of source attribution and their characteristics

Type of source attribution	Data source	Characteristics and challenges
Case-control	Sporadic cases	Confounding (e.g. likely to be part of a cluster; attributing the disease to a factor while actually being related to another correlated factor)
Outbreak investigation	Outbreaks	Based on epidemiological, food chain and microbiologic data, uncertainty
Subtyping models	Cases related to food	Uncertainty, lacking relation between types in patients and not in foods
Quantitative microbiological risk assessment (QMRA)	Prediction of cases	Large variability, uncertainty, fail-safe assumptions, often overestimation of number of cases
Expert elicitation	Experts	Biases, uncertainty
Natural experiments	Sudden changes in consumption related to public health effects	Confounding, only useable in very special situations

In comparison to many other foodborne diseases, the specificities of listeriosis source attribution make the uncertainty smaller, because the proportion of food-related cases is very large (close to 100 percent). Furthermore, the case definition for listeriosis is well described, and detection, identification and genotyping are often performed; therefore, there is relatively low under-reporting (due to the severity).

3.2 FOOD ATTRIBUTION OF LISTERIOSIS

Pires *et al.* (2020) describe the food attribution of listeriosis as being 100 percent, while Havelaar *et al.* (2012) estimated it as being 69 percent. Cressey *et al.* (2019) estimated the food attribution at 88 percent and compared it with various other reported data (including the Havelaar *et al.* estimate) that is, 85, 100, 99, 77, 98, 99 and 69 percent. Using this latter data, the average estimate would be 89 percent (± 12 percent), with a range of 69 to 100 percent.

3.3 FOOD SOURCE ATTRIBUTION

Various literature sources were compared to estimate the attribution of listeriosis to specific food groups. A number of different food group descriptors have been used by researchers. Some had deli meat and frankfurters, where others separated meat into pork, beef and chicken. Therefore, only a general source attribution was performed, and results from the literature were combined into the groups meat, dairy, (shell)fish, fruits and vegetables, other and non-food. Data are presented in Table A4 in Annex 4.

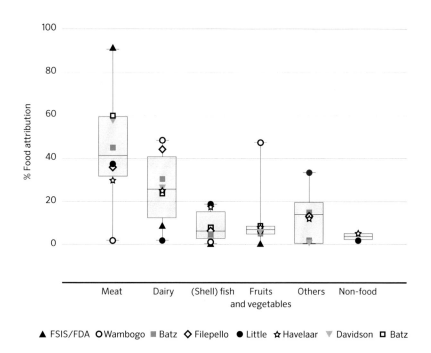

FIGURE 3. Food attribution of listeriosis compiled from various references

Source: Adapted from Batz *et al.*, 2012, 2014; Davidson *et al.*, 2011; FDA/FSIS. 2003; Filipello *et al.*, 2020; Havelaar *et al.*, 2012; Little *et al.*, 2010; Wambogo *et al.*, 2020.

3.4 CONCLUSIONS

Several (national) listeriosis source attribution studies have been reported using various methods (e.g. expert opinion, genomic data analysis, risk assessment). These studies are, however, limited to high-income regions, and no global listeriosis source attribution study exists to date. Results show large variation; however, generally, it can be concluded that meat and dairy are the most frequent sources.

For source attribution studies, it would be useful if harmonized food classification schemes (ontology) were used at the international level for the categorization/typing of foods considered. This is complicated by the fact that for certain pathogens the specific source is more relevant (*Campylobacter*/poultry), while for others the consumption type is more relevant (*Listeria*/deli meat/RTE).

Further source attribution studies (based on systematic reviews of outbreak data, case-control data of sporadic cases, and genomic analysis) at the global level would provide better insight into the quantitative relevance of sources and can help in targeting management options.

4

Host susceptibility

Listeriosis is a severe disease that mainly affects high-risk subpopulations (EFSA and ECDC, 2017; Buchanan *et al.*, 2017; Charlier *et al.*, 2017; Friesema *et al.*, 2015). Listeriosis is subdivided clinically into non-invasive (cutaneous listeriosis, endophthalmitis, gastroenteritis) and invasive forms. In the present document, only invasive listeriosis is discussed, and it is classified into three forms: bacteraemia, neurolisteriosis (meningitis, encephalitis, rhomboencephalitis), and maternal-neonatal infection.

4.1   CLASSIFICATION OF SUBPOPULATIONS

The identification of different at-risk groups of consumers can be very helpful to inform health risk analysis strategies. For example, individuals at greater risk of acquiring listeriosis often have a low awareness of the potential risks of foodborne listeriosis, but this awareness can play a key role in prevention, as can an individual's attendant/caregiver (Maia *et al.*, 2019). Thus, rather than recommending the potential revision of the microbiological criteria for defined subpopulations, the expert group recommends that public risk communication should also be focused on informing identified at-risk subpopulations and their attendants/caregivers about their relative susceptibility, as well as about the foods that have a relatively high probability of containing *L. monocytogenes* and causing foodborne listeriosis (Goulet *et al.*, 2012a, 2012b; Maia *et al.*, 2019).

This variability of susceptibility in a population could be divided into three sub-populations (FDA/FSIS, 2003), namely less susceptible, susceptible and very susceptible.

4.1.1 Less susceptible subpopulations

Less susceptible subpopulations, also called "healthy individuals/adults", are defined as the general population (non-pregnant persons) under 65 years old of age, with no known underlying conditions that would predispose someone to being at greater risk for acquiring foodborne listeriosis.

This less susceptible subpopulation is difficult to define because:
  (i)   some patients discover their underlying condition(s) after infection;
  (ii) a small portion of the population are healthy carriers of *L. monocytogenes* without a known status, and no updated estimation of this exists in the "healthy individuals" category (Painter and Slutsker, 2007);
  (iii) individuals without any known risk factors for listeriosis have occasionally become severely infected (FAO and WHO, 2004b). Unknown individual predisposition of their host immune systems and perhaps host genetics or intestinal microbiomes could increase the risk of listeriosis; and
  (iv) the non-invasive forms of listeriosis, mainly gastroenteritis, without progression to invasive forms are underestimated and not well known (Ooi and Lorber, 2005).

Based on 1) insufficient data available to estimate the public health impact of gastroenteritis, and 2) the minor relevance of GI listeriosis to the overall burden of the disease, the non-invasive form of listeriosis was not considered in the current exercise and is not recommended to be included if a new risk assessment of *L. monocytogenes* in RTE food is done.

4.1.2 Susceptible subpopulation

Susceptible subpopulations include pregnant women and their newborns (neonates: under 28 days old) and adults aged 65 or older (Buchanan *et al.*, 2017).

During pregnancy, listeriosis occurs following the consumption of a variety of high-risk foods that can support the growth of *L. monocytogenes*. For elderly individuals, the factors contributing to an increased risk of acquiring listeriosis appear to be improper food storage and handling practices as well as the consumption of food after the shelf-life date (Maia *et al.*, 2019).

4.1.3   Very susceptible subpopulation

A very susceptible subpopulation is defined as people with weakened immune systems, such as individuals with untreated HIV infection (CD4$^+$ T lymphocyte count < 200 cells/mm^3), cancer patients, organ transplant patients and elderly and/or immunosuppressed patients. For these last patients, risk managers in the European Union (EU) developed a specific microbiological criterion to protect these consumers: "No detection of *L. monocytogenes* in 10 samples of 25 g of ready-to-eat foods intended for infants and for special medical purposes placed on the market during their shelf-life" (EC No 2073/2005) (Falk *et al.*, 2016; Goulet *et al.*, 2012a, 2012b).

4.2   RELATIONSHIPS WITH DOSE RESPONSE

Since *L. monocytogenes* can contaminate and grow in a wide range of foods, many people ingest small numbers of this pathogen quite frequently without showing any symptoms.

Based on the knowledge that listeriosis is dose dependent (even if the low dose impact should be better investigated [Holcomb *et al.*, 1999]), statistical models supported by animal studies have rapidly evolved by using gerbils or humanized mice, and more accurate human outbreak data has been used to better quantify the dose-response relationship (Rahman *et al.*, 2018). The relationship between the dose ingested and the likelihood of severe listeriosis (response) depends on the immune status of the host and the virulence of the *L. monocytogenes* strain. In the past, epidemiological data combined with dose-response models have strongly suggested that the ingestion of a low dose of *L. monocytogenes* leads to an average low probability of invasive listeriosis in the general population, as well as in broadly defined populations with heightened susceptibility (Chen *et al.*, 2003; FAO and WHO, 2004b).

In 2004, FAO/WHO experts developed an exponential dose-response model for invasive listeriosis (FAO and WHO, 2004b; Rocourt *et al.*, 2003) that assumes:
- the probability of a given bacterial cell causing the adverse effect is independent of the number or characteristics of other ingested pathogens, so that a single ingested microorganism is sufficient to cause an adverse effect with some probability greater than zero; and
- the bacterial cells are randomly distributed in the food and the average probability *r* that one cell, with a given exposure of a particular consumer to a specific population of pathogens, will survive to the host-pathogen interaction to initiate infection and cause illness is constant.

The exponential dose-response model for listeriosis does not take into account the effects of strain variation and individual/subgroup susceptibility on dose-response outcomes. However, Pouillot *et al.* (2015) did develop an exponential dose-response model incorporating adjustments for variability in *L. monocytogenes* strain virulence and host susceptibility. In addition, a new mechanistic dose-response model of *L. monocytogenes* infection in human populations which also takes into account the effects of strain variation and individual/subgroup susceptibility was recently described by Rahman *et al.* (2018).

According to this recent model (Pouillot *et al.*, 2015), the marginal probability of developing invasive listeriosis upon ingestion of one cell of *L. monocytogenes* per individual for the general population is 8×10^{-12} and 3×10^{-9} for extremely susceptible subpopulations. The probability of acquiring listeriosis is at least 100 times greater with the most virulent strains, belonging to the so-called hypervirulent clones (Pouillot *et al.,* 2012).

The possibility of exploring the development of different dose-response models should be investigated. One approach could be to base it on the susceptibility of the population and to develop different dose-response models for each of the three main subpopulations to adequately characterize the listeriosis risk: (i) less susceptible subpopulation, (ii) susceptible subpopulation, and (iii) very susceptible subpopulation. These different dose-response models for each of the three main subpopulations could then become a key tool for informing risk assessments and evaluating different risk management strategies for the control of *L. monocytogenes* in RTE foods (Buchanan *et al.*, 2017). In general, there is a need for new data and insights for *L. monocytogenes* dose-response models (Chen *et al.*, 2011; Hoelzer *et al.*, 2012a, 2013). Revisiting dose-response models taking into account new animal models and specific clonal complexes or sublineages could provide new insights and could be very instructive in comparison to using reference strains.

4.2.1 Risk factors and comorbidities

A literature search was conducted to identify prevalence estimates of risk factors since they drive the vulnerability to listeriosis (Falk *et al.*, 2016; Brent, 2012; Dalton *et al.*, 2011; Fernàndez-Sabé *et al.*, 2009; Friesema *et al.*, 2015; Gerner-Smidt *et al.*, 2005; Mook *et al.*, 2012; Preussel *et al.*, 2016; Rocourt, 1996; Scobie *et al.*, 2019). It should be emphasized that all comorbidities are not risk factors, as only comorbidities that affect the immune system should be called "risk factors". An underlying condition such as being elderly, pregnancy, cancer and immunosuppressive therapy and untreated AIDS, which predisposes one to listeriosis by interfering with CD4[+] T-cell mediated immunity response, was observed in a large majority of patients. Ten risk factors could be defined (see Table 3).

According to Goulet *et al.* (2012a), those with chronic lymphocytic leukaemia had a > 1 000-fold increased risk of acquiring listeriosis, while those with liver cancer; myeloproliferative disorder; multiple myeloma; acute leukaemia; giant cell arteritis; dialysis; oesophageal, stomach, pancreas, lung, and brain cancer; cirrhosis; organ transplantation; and pregnancy, had a 100–1 000-fold increased risk of listeriosis as compared with French persons < 65 years of age with no underlying conditions.

In the work of Pouillot *et al.* (2015), who revisited the exponential dose-response curve for invasive listeriosis, the subpopulation was not based on group susceptibility, but rather on ten more precisely defined subpopulations with similar underlying conditions based on underlying pathophysiology and the expected degree of CD4[+] T-cell inhibition, according to the classification of risk factors by Goulet *et al.* (2012a) (Table 3).

TABLE 3. Subpopulation descriptions and their corresponding relative risk values for cases of invasive listeriosis in France between 2001 and 2008 and resulting statistics for r, the probability of illness following the ingestion of one cell of *L. monocytogenes* of the lognormal-Poisson dose-response model for invasive listeriosis

Subpopulation	Description	Relative Risk (CI 95%)[a]	Estimates of r : mean[b] Lognormal Poisson dose-response model
Less than 65 years old, no known underlying condition, healthy adults	Population < 65 years with no conditions	Reference group	7.90×10^{-12}
Diabetes	Type I, Type II	7.6 (3.5, 15.6)	7.47×10^{-11}
Heart disease	Self-reported heart disease	5.4 (1.5, 14.4)	5.01×10^{-11}
Inflammatory disease	Rheumatoid arthritis, Crohn's disease, colitis, ulcerative colitis, giant cell arteritis	58.5 (25.2, 123.4)	8.43×10^{-10}
Cancer (non-haematological)	Breast, brain, ear, nose and throat, gastrointestinal, gynaecological, kidney, liver, lung, prostate cancers	54.8 (34.2, 90.3)	7.76×10^{-10}
HIV/AIDS	HIV or HIV/AIDS	47.4 (10.5, 140.4)	6.50×10^{-10}

(cont.)

More than 65 years old, no known underlying condition	Population ≥ 65 years with no conditions	13.9 (8.6, 23.1)	1.49×10^{-10}
Cancer (haematological)	Leukaemia, Hodgkin's lymphoma, non-Hodgkin's lymphoma, multiple myeloma	373.6 (217.3, 648.9)	9.60×10^{-9}
Solid organ transplant	Heart, intestinal, kidney, liver, lung, and pancreas transplant patients	163.7 (26.3, 551.5)	3.14×10^{-9}
Renal or liver failure	Dialysis: haemodialysis, peritoneal dialysis, liver disease: hepatitis A, B, C	149.4 (82, 270.1)	2.79×10^{-9}
Pregnancy[c]	Total number of live births + foetal loss + abortions/ population x 0.75	116 (71, 194.4)	2.01×10^{-9}

Source: Adapted from Falk *et al.*, 2016; Goulet *et al.*, 2012a; Pouillot *et al.*, 2015.

[a] Estimated using a Poisson regression without adjustment. These 95 percent CIs should be considered only as indicative but suggest that all those groups have a risk of listeriosis significantly greater than the reference group.

[b] Resulting statistics for *r*, the probability of illness following the ingestion of one cell of *L. monocytogenes* obtained from the Lognormal-Poisson dose-response model for invasive listeriosis following the ingestion of *L. monocytogenes* in different population subgroups. The distribution of *r* includes the individual within group and the strain variability (Pouillot *et al.*, 2015).

[c] Prevalence of pregnancy was determined using the Statistics Canada CANSIM database and a report from the Canadian Institute for Health Information. Estimates were determined by first summing live births, foetal loss and abortions for 2011. This value was then divided by the 2011 population. To account for the length of pregnancy, nine months, the value was finally multiplied by 0.75. National, provincial and territorial estimates were calculated. (For further detail, see Supplementary Material of Falk *et al.*(2016), available at: https://www.cambridge.org/core/journals/epidemiology-and-infection/article/comparing-listeriosis-risks-in-atrisk-populations-using-a-userfriendly-quantitative-microbial-risk-assessment-tool-and-epidemiological-data/FBC5713ED1BE3F62A0B BEE47F762332C#supplementary-materials).

Summing prevalence estimates across risk factors has been used to estimate the proportion of susceptible individuals in previous listeriosis risk assessments, but this approach has drawbacks in that it does not take into account comorbidities or the fact that more than one risk factor could be present in individuals with listeriosis (Falk *et al.*, 2016). The model also does not explicitly incorporate additive or synergistic risk for those individuals with multiple risk factors (Falk *et al.*, 2016).

In general, the *L. monocytogenes* quantitative microbiological risk assessment (QMRA) models have several limitations. For example, region-specific risk factors and relative risk values were sometimes unavailable. The studies of Goulet *et al.* (2012a) and Pouillot *et al.* (2012) were originally used as the basis to derive the risk factors and relative risk estimates, but their generalization should be investigated based on general consumption patterns in vulnerable subpopulations that could influence the relative risk factors derived from notified listeriosis cases (Falk *et al.*, 2016). The subpopulations characterized by Goulet *et al.* (2012a) have evolved since 2008, and although they are still relevant, it would be beneficial, if feasible, to redefine them with updated data to better inform the dose-response models.

For the accuracy of data, the country selected for this update of subpopulations and their risk values should be one that has an extensive recording of listeriosis cases based on mandatory notification, all of which should be validated by a capture-recapture study – a statistical technique used in infectious disease surveillance to estimate the completeness of the notification of the number of patients with an infectious disease (Goulet *et al.*, 2012b). New approaches for dose-response models using subpopulations and the virulence of strains implies that the selected countries should have a large collection of well-characterized human listeriosis strains. The last extensive cohort study, even if it was more precise in the analysis of clinical data, did not cover a more recent period, so it could not be of immediate help to more accurately update the data concerning the risk factors of different subpopulations (Charlier *et al.*, 2017).

The absence of data on other risk factors for listeriosis such as alcoholism, antacid use, corticosteroid therapy and laxative use has been identified and could have a significant impact on future dose-response models (Charlier *et al.*, 2017; Falk *et al.*, 2016). More research on listeriosis risk factors at the international level, their overlap or synergy, and their resulting contributions to the risk of listeriosis should be emphasized.

The expert group recognizes the need for a future update on the relative risk values of risk factors and comorbidities used in dose-response models for listeriosis, based on recent large cohort studies and newly identified potential risk factors such as a high-fat diet (Las Heras *et al.*, 2019; Shinomiya *et al.*, 1988) and so on. This data needs to be collected from the new estimated global burden of listeriosis.

For the severity of listeriosis and general risk factors, please also see Sections 7.2 and 7.3.

5

Current monitoring, surveillance and control programme

5.1   INTRODUCTION

The FAO/WHO risk assessment for *L. monocytogenes* in RTE foods (FAO and WHO, 2004a, 2004b) provided scientific advice that served as the foundation for the subsequent development of Codex guidelines (FAO and WHO, 2009) for the control and management of *L. monocytogenes* in foods.

The guidelines provided advice to governments and the food industry on a framework to minimize the likelihood of illness arising from the presence of *L. monocytogenes* in RTE foods. It included microbiological criteria for *L. monocytogenes* in RTE foods and recommendations for the establishment of environmental monitoring programme for *L. monocytogenes* in food-processing facilities.

Recent research and data derived from different food commodities and geographical regions provides the impetus to consider the application of these guidelines to a wider range of food commodities and consider new management approaches to control *L. monocytogenes*. Specifically, this necessitates a review of current monitoring and assurance programme for establishing the presence of *L. monocytogenes* in food production and food-processing environments, and in final products, and their role in management and control of this pathogen.

Scientific evidence has demonstrated that most effort should focus on the management and control of the hazards in a more proactive way by implementing an effective food safety management system (FSMS). Recommendations included in the Codex guidelines highlight the necessity for an environmental monitoring programme, and there is an increasing emphasis on environmental monitoring to assess incursions and persistence of *L. monocytogenes* into processing facilities and onto food-contact surfaces and its persistence in these environments.

Typically, such monitoring aims to find *Listeria* spp. (which includes *L. monocytogenes*) in the food-processing environment, with some regulatory authorities requiring further testing to specify positive findings. Increasingly these authorities require whole genome sequencing (WGS) of *L. monocytogenes* isolates, as this assists with tracking and tracing outbreaks, identifying virulence factors, and determining whether strains have established residence in food-processing facilities.

However, in some countries, subtyping of pathogenic isolates is not mandatory, and different countries do it only on a voluntary basis. This generates a data deficiency problem which can be overcome by increasing the level of support from organizations such as FAO/WHO, as well as harmonization at the national and international level. Another issue is the capacity and resources of low-to-middle income economies to be able to undertake WGS.

Alternatively, testing for other indicator organisms such as the *Enterobacteriaceae* or screening for adenosine triphosphate (ATP) may be used to assess the efficacy of cleaning and sanitation programme. For this reason, regulatory agencies should encourage aggressive environmental monitoring to eliminate sources of *L. monocytogenes* (Farber *et al.*, 2021).

Ineffective sampling programme or sampling techniques are a concern as they may result in the non-detection of *L. monocytogenes* when it is present. This would prevent the implementation of corrective actions and give a false sense of security (Lahou and Uyttendaele, 2014).

Investigation of outbreaks of foodborne illness have contributed much to understanding the importance of acquired characteristics of known pathogens, prioritization of emerging pathogens, rapid detection of known pathogens in routine monitoring, timely national/international communication and cooperation, and most importantly application of a proactive approach in foodborne illness surveillance systems (Yeni *et al.*, 2017).

In order to reduce foodborne listeriosis, risk communication strategies must also be developed to clearly communicate risk factors associated with product storage, shelf-life and appropriate consumption of RTE foods by susceptible and very susceptible consumers.

Fresh produce has emerged as an important source of foodborne illness outbreaks linked to *L. monocytogenes,* and this has become a global public health problem, partly due to international trade of these products. It is not clear if consumers recognize the link between consumption of fresh produce and outbreaks of listeriosis. Additional risk communication in this area is recommended. This is increasingly important for produce that doesn't receive a kill-step which is able to eliminate *L. monocytogenes* from the finished product (non-detected in 25 g), or for food products which are consumed in their raw state.

In the case of frozen non-RTE vegetables, adherence to package instructions for cooking at recommended temperatures prior to consumption is important, and consistency in label instructions may improve adherence to recommended guidance.

5.2 SCOPE

RTE food is defined as any food which is normally eaten in its raw state, or any food handled, processed, mixed, cooked, or otherwise prepared into a form which is normally eaten without further listericidal steps (FAO and WHO, 2009).

This document seeks to review current monitoring strategies for establishing the presence of *L. monocytogenes* in RTE foods, food production and processing environments.

It includes the tabulation of data on monitoring and surveillance received from member states. It reaffirms definitions for key terms (monitoring, surveillance, verification and validation) in relation to their role in managing the risk presented by *L. monocytogenes* in food. These definitions are taken from relevant Codex documents and WHO sources.

It includes consideration of the microbiological criteria in Annex II of the Codex guidelines on the application of general principles of food hygiene to the control of *Listeria monocytogenes* in food (FAO and WHO, 2009), which establishes limits on the basis of whether a food will or will not support the growth of *L. monocytogenes*.

At this stage, only a few countries such as the United States of America and Türkiye require non-detection of *L. monocytogenes* in 25 g of foods (referred to as zero-tolerance) and thereby still have a "zero tolerance" approach for *L. monocytogenes* for all RTE foods.

Attention is also being paid to the growing range of foods implicated in outbreaks of listeriosis and to the challenge presented by the consumption of foods which are not typically considered or intended to be RTE. It is important to understand that consumers also have a role in food safety. Although outside current food safety legislation, consumers should ensure that food is stored, handled and prepared in a manner that ensures it is safe for consumption (EFSA, 2020). Retailers and producers should provide the consumer with information to ensure food safety during storage, handling and preparation of the product, for example using a leaflet or by verbally informing the consumer (EFSA, 2017).

5.3 MONITORING AND SURVEILLANCE PROGRAMME

The aim of this section is to provide an overview of data from different regions and countries, as well as peer reviewed and grey literature on monitoring and surveillance activities for *Listeria* spp. and *L. monocytogenes* in foodstuffs and food-processing environments.

This data reported reflects country-specific testing programme, which have targeted selected commodities under policies and guidelines issued by competent authorities (CA).

Monitor: The act of conducting a planned sequence of observations or measurements of control parameters to assess whether a control measure is under control (FAO and WHO, 2020).

Verification: The application of methods, procedures, tests and other evaluations, in addition to monitoring, to determine whether a control measure is or has been operating as intended (FAO and WHO, 2020).

Validation (control measures): Obtaining evidence that a control measure or combination of control measures, if properly implemented, is capable of controlling the hazard to a specified outcome (FAO and WHO, 2020).

Surveillance: This is a term which is often used within food safety, public health, and wider global health systems, and its meaning varies depending on the situational context. The definition of surveillance in this report is that used by FAO within the context of a national food safety system. Therefore, surveillance means the systematic ongoing collection, collation and analysis of information related to food safety and the timely dissemination of information for assessment and response as necessary (FAO and WHO, 2019). For other definitions of "Surveillance", please see FAO, 2007 and EFSA, 2020.[1]

5.3.1 Regulatory limits (Codex guidelines and approaches used in different countries and regions)

Current Codex guidelines (FAO and WHO, 2009) include microbiological criteria for specific categories of RTE foods which are intended as advice to governments within a framework for the control of *L. monocytogenes* in RTE foods, with a view towards protecting the health of consumers and ensuring fair practices in food trade.

The different approaches that can be found in the different countries can be divided into two groups or international standards. Firstly, the European Union and many other countries have adopted a risk-ranking system based on end product type, use and challenge testing. Secondly, the United States of America and Türkiye have adopted the philosophy of non-detection of *L. monocytogenes* in 25 g, which has been defined as "zero tolerance" for *L. monocytogenes* in the processing environment and RTE foods.

In the United States of America, a product is considered RTE if there is a standard of identity (e.g. hotdogs or barbeque) or a common or usual identity (e.g. pâtés) defining the product as fully cooked, or if it meets the definition in the *Listeria* Rule (9 CFR 430.1). Examples of RTE products include deli products, hot dog products, whole hams, sausages, meat salads, and other products that have been treated with a lethality step (FSIS and USDA, 2014). Fresh produce is also included and, in 2017, a "zero tolerance" standard for *L. monocytogenes* was established for sprouts. On the other hand, the Australia New Zealand Food Standards Code (FSC) definition for RTE food excludes shelf-stable foods, whole raw fruits, whole raw vegetables, nuts in the shell and live bivalve mollusks.

Most of the standards developed by CA provide guidance regarding growth of *L. monocytogenes* in RTE foods – growth will not occur if:

---

[1] It is defined as "*An official process which collects and records data on pest occurrence or absence by survey, monitoring or other procedures*", in ISPM No. 5 (FAO, 2007), and "*Surveillance sampling (product sampling) to seek for the prevalence of pathogens (e.g. official monitoring of CAs, food products on the market)*" in EFSA (EFSA, 2020).

(a) the food has a pH less than 4.4 regardless of water activity; or

(b) the food has a water activity less than 0.92 regardless of pH; or

(c) the food has a pH less than 5.0 in combination with a water activity of less than 0.94; or

(d) the food has a refrigerated shelf-life no greater than 5 days; or

(e) the food is frozen (including foods consumed frozen and those intended to be thawed immediately before consumption); or

(f) it can be validated that the level of *L. monocytogenes* will not increase by greater than 0.5 log CFU/g over the food's stated shelf-life (Dairy Authority of South Australia, 2015).

In those cases where a risk-ranking system is based on end product type and its usage, challenge testing is applied, and the food is divided into two groups, with some exceptions:

- products in which growth will not occur, have a limit of up to 100 CFU/g in the product (5 samples × 25g); and

- products in which growth may occur, have a limit of non-detection in 25 g (non-detected in 25 g, 5 samples).

In Europe, as an example, in those cases where the RTE food supports growth of the pathogen, but during its shelf-life the levels of *L. monocytogenes* do not exceed 100 CFU/g, the food business operator (FBO) may fix intermediate limits (at the end of the production process) that should be low enough to guarantee that the limit of 100 CFU/g is not exceeded at the moment of consumption without cooking (that is to say, as an RTE food) (EC regulation 2019/229) (EFSA, 2020).

These intermediate limits should be established taking into account the potential growth of the pathogen during storage for a certain period under temperature conditions allowing for its growth. The reasonably foreseeable conditions of use by the consumers need to be considered beyond the recommendations provided by the FBO through the product labelling.

An approach can be to determine acceptable *L. monocytogenes* concentration (CFU/g) that could be considered as a performance objective (PO) at the end of the production process, immediately before releasing the RTE product on the market, compatible with the food safety objective (FSO) of 100 CFU/g.

In the United States of America, two agencies are involved with food safety: the United States Department of Agriculture Food Safety and Inspection Service (FSIS) and the United States Food and Drug Administration (FDA).

The FSIS *Listeria* Rule (9 CFR part 430; Control of *L. monocytogenes* in RTE meat and poultry products) describes three alternative methods establishments can use in controlling *L. monocytogenes* contamination of post-lethality exposed RTE meat and poultry products:

- **Alternative 1**: An establishment applies a post-lethality treatment (PLT) to reduce or eliminate *L. monocytogenes* **and** an antimicrobial agent or process (AMAP) to suppress or limit growth of *L. monocytogenes*.
- **Alternative 2**: An establishment applies either a PLT or an AMAP.
- **Alternative 3**: The establishment does not apply any PLT, AMAP; instead, it relies on its sanitation programme to control *L. monocytogenes*.

Independent of the strategy implemented by the industry, companies are required to validate that their processing systems can produce compliant produce. They are also expected to verify that their processing environment and final product complies with the regulations. These verification records are reviewed by the CA.

5.3.2 Food classification between foods that support growth and foods that do not support growth

The growth and survival of *L. monocytogenes* is influenced by many factors. In food, these include temperature, pH, water activity, salt and the presence of preservatives (Table 4), as well as the background microflora. The temperature range for growth of *L. monocytogenes* is between -1.8 °C (ca. -2° C) and 45 °C, with the optimal temperature being 30–37 °C. Temperatures above 50 °C are lethal to *L. monocytogenes*. Freezing can lead to a small reduction in *L. monocytogenes* numbers (Lado and Yousef, 2007). *L. monocytogenes* will grow in a broad pH range of 4.0–9.6 (Lado and Yousef, 2007).

The ranges of environmental factors that permit growth of *L. monocytogenes* are discussed in detail in a number of reviews and have been summarized in the FAO/WHO technical report (FAO and WHO, 2004b). Table 4 shows a summary of these ranges; however, it should be noted that these limits are not absolute (FAO and WHO, 2004b).

TABLE 4. Ranges of environmental factors that permit growth of *L. monocytogenes* when all other factors are optimal

Environmental factor	Limits	
	Lower limit	Upper limit
Temperature (°C)	-2 to +4	~ 45
Salt (% water phase NaCl) (and corresponding aw)	< 0.5 (0.91–0.93)	13–16 (> 0.997)
pH (HCl as acidulant)	4.2–4.3	9.4–9.5
Lactic acid (water phase)	0	3.8–4.6 mM, MIC[1] of undissociated acid[2] (800–1 000 mM, MIC of sodium lactate[3])
Acetic acid	0	~20 mM (MIC of undissociated acid)
Citric acid	0	~3 mM (MIC of undissociated acid)
Sodium nitrite	0	8.4–14.4 µM (undissociated)

Sources: The overall ranges are summarized from Ryser and Marth, 1991; ICMSF, 1996; and Augustin and Carlier, 2000

Notes: [1] MIC = minimum inhibitory concentration, i.e. the minimum concentration that prevents growth. [2] From Tienungoon, 1998. [3] From Houtsma, de Wit and Rombouts, 1993 (FAO and WHO, 2004b).

5.3.3 Establishment of suitable indicators for *Listeria monocytogenes*

Relationship between Listeria *spp. and* L. monocytogenes

Although *L. monocytogenes* is the only human pathogen within the genus *Listeria* (except for a few human cases that have been caused by *L. ivanovii*), food businesses often test for *Listeria* spp. instead of specifically focusing on *L. monocytogenes*. This is both more rapid and cost effective and provides evidence of conditions likely to support the survival and growth of *L. monocytogenes* if it is present. *Listeria* spp. can survive and multiply under adverse conditions, and their detection is a good indicator of inadequate hygiene or cleaning and sanitation of food handling areas.

The prevalence of *L. monocytogenes* correlates well with that of other *Listeria* spp. for some but not all food-processing operations. Facilites producing RTE meat products are characterized by a varied prevalence of *Listeria* spp. with inconsistent correlation between contamination by *L. monocytogenes* and other *Listeria* spp. The presence of *Listeria* spp. does not seem to be a consistent universal indicator for *L. monocytogenes* prevalence in seafood-processing facilities (Alali *et al.*, 2013). There are also large variations amongst strains of *L. monocytogenes* in their competitiveness under multibacterial culture conditions, with specific *L. innocua* strains capable of inhibiting the growth of *L. monocytogenes*, an action that can be further enhanced by the presence of a diverse background of gram-negative bacteria. Both *Listeria* selective agar and PALCAM agar display a low sensitivity and specificity in *L. monocytogenes* detection as compared to CHROMagar™ Listeria which demonstrated a 96.9 percent and 99.1 percent sensitivity and specificity, respectively, for *L. monocytogenes* detection in naturally contaminated foods (Jamali, Chai and Thong, 2013).

Other indicators

In the case of general process hygiene indicator microorganisms, the presence and/or concentration of these indicator organisms should reflect states or conditions that indicate process control or lack of process control. Flexibility should be provided so that the most effective verification systems can be implemented at the establishment level. The most common and applicable are listed here.

Enterobacteriaceae

Enterobacteriaceae counts reflect, in addition to faecal contamination, the level of environmental hygiene. *Enterobacteriaceae* are usually considered by food manufacturers as hygiene indicators and are therefore used to monitor the effectiveness of implemented preventive prerequisite measures such as Good Manufacturing Practices and Good Hygiene Practices (GMP/GHP) (Cox, Keller and Van Schothorst, 1988). This is also reflected in numerous national and international standards or criteria where *Enterobacteriaceae* or coliforms are included as hygiene indicators with 3-class sampling plans. *Enterobacteriaceae* testing serves the same purpose as coliform testing in that it indicates improper cleaning, unsanitary conditions and post-process contamination (3M, 2021).

Aerobic Plate Count (Total Viable Count)

The Aerobic Plate Count (APC) can be regarded as being a reliable indicator of the overall level of bacterial contamination in the environment and food sample. It provides information on the total population of bacteria present. APC counts above a certain threshold would typically suggest that sanitation of the specific environment or equipment was ineffective or improperly performed (3M, 2021).

Escherichia coli

Generic *E. coli* is a good indicator of faecal contamination. It has been used for many years as an indicator of faecal contamination in water treatment, because it is present in almost all faecal samples. Generic *E. coli* is generally considered a better indicator of the potential for faecal contact than APC or coliforms but does not necessarily indicate the presence of pathogens (Belias *et al.*, 2021).

Alternative approaches to microbiological indicators

Alternative approaches to microbiological testing that are properly validated should be established where they offer practical advantages such as the ATP+ADP+AMP test. ATP hygiene monitoring tests are widely used for assessing the effectiveness of cleaning procedures. The test is easy to use and gives immediate results; however, ATP can be metabolized and degraded to ADP and AMP. Recently, a total adenylate [ATP + ADP + AMP] monitoring test has been developed for cleaning verification in healthcare settings (Bakke *et al.*, 2019).

5.3.4 Relationship between regulatory limits and listeriosis incidence

The European Food Safety Authority (EFSA) scientific opinion (EFSA, 2018) concluded that despite the application of the food safety criteria (FSC) for *L. monocytogenes* in RTE foods from 2006 onwards (Commission Regulation (EC) 2073/2005), a statistically significant increasing trend of human invasive listeriosis was reported in the European Union and European Economic Area (EU/EEA) over the period from 2009 to 2013 (EFSA and ECDC, 2015). In 2010–2011, an European Union-wide baseline survey (BLS) estimated the prevalence and concentration of *L. monocytogenes* in RTE foods at retail: packaged (not frozen) smoked or gravad fish, packaged heat-treated meat products and soft or semi-soft cheese. Based on this report, the European Union-level estimate of the proportion of samples with *L. monocytogenes* counts > 100 CFU/g at the end of shelf-life was 1.7 percent for "RTE fish," 0.43 percent for "RTE meat" and 0.06 percent for "RTE cheese."

Improved control measures starting in the 1990s have greatly reduced the prevalence of *L. monocytogenes* in many food categories, particularly in RTE meats and meat products. However, the rate of illness has remained constant during the last decade. Furthermore, recent outbreaks have challenged the conclusions of existing risk assessments and our understanding of the influence of virulence, host and food matrix on foodborne illness (EFSA and ECDC, 2018a; Buchanan *et al.*, 2017).

In the case of the United States of America, the establishment of improved control measures started in the 1990s. These control measures have greatly reduced the prevalence of *L. monocytogenes* in many food categories, particularly in RTE meats and meat products. Despite these improvements, the rate of listeriosis has remained constant during the last decade and the more severe, systemic (invasive) form of listeriosis is now recognized as occurring more frequently in small outbreaks than previously recognized (Buchanan *et al.*, 2017). The use of WGS in listeriosis outbreak investigations is also an important factor in why the United States of America is seeing a larger number of small outbreaks.

Other considerations of monitoring and surveillance programme

How can one define the "relevant amount of growth" in a food category classified as high risk?

No growth does not really mean zero growth, as there is always a limit of detection for a method. However, mostly due to the variability of the microbial enumeration techniques, it has been generally accepted that increases below a 0.5 log growth over the shelf-life of a food are considered as no growth.

The growth potential (δ) is defined as the difference between the $\log_{10}$ CFU/g at the end of the test and the $\log_{10}$ CFU/g at the beginning of the test (Beaufort *et al.*, 2014). A food is considered capable of supporting the growth of *L. monocytogenes* if the δ is higher than 0.5 $\log_{10}$ CFU/g, while it is assumed that the food is not able to support the growth if the δ is lower than 0.5 $\log_{10}$ CFU/g.

Based on the QMRA model used by EFSA (EFSA, 2018) in the scientific opinion of "*L. monocytogenes* contamination of RTE foods and the risk for human health in the European Union", it was found that 92 percent of invasive listeriosis cases for all age-gender groups are attributable to doses above 10^5 CFU per serving. Assuming an average serving size of 50 g, this would correspond to an average *L. monocytogenes* concentration in RTE foods above 2 000 CFU/g at the time of consumption. However, a small proportion of cases may be associated with those RTE foods that have lower *L. monocytogenes* levels. Results from the QMRA model indicated that differences in consumption among the age groups influenced the probability of exposure to *L. monocytogenes* through the effect on the prevalence. This suggests that part of the increase in invasive listeriosis incidence with age can be explained by consumption, i.e. the overall prevalence of *L. monocytogenes* in the generic RTE food weighted to reflect consumption increases with ages over 25–44 years.

Currently, there is a wide disparity among member countries in terms of the awareness, knowledge, sophistication and funding of their current monitoring and surveillance programme for human listeriosis. This ranges from countries having a well-funded, active surveillance programme for listeriosis with sentinel sites set up across the country, to those that have either no surveillance programme or a rudimentary, passive surveillance programme, where the actual burden of listeriosis is likely vastly underestimated due to a lack of both specific surveillance and standard mandatory reporting of listeriosis cases. The surveillance programme from various countries as discussed at this JEMRA meeting can be found in Annex 3.

When an RTE food is found to exceed microbiological limits or the presence of *L. monocytogenes* on food contact surfaces is confirmed, it could be a reason to initiate a recall. The CA in some jurisdictions actively monitors for the presence of *L. monocytogenes* in RTE foods and/or environmental samples, but other competent authorities (CAs) use an auditing process to provide oversight on any monitoring done by a company. As such, there is a diversity in the approach taken by different countries, with some using a more command and control system while others use a more outcome-based approach (For further details, see Annex 3).

Shelf-life labelling

Newsome *et al.* (2014) examined applications and perceptions of date labelling of food from a global perspective. The many variations in date labelling (e.g. use by, consume by, best before, expires on) contribute to confusion and misunderstanding regarding how the dates on labels relate to food quality or safety. These latter issues imply that date labelling of foods is likely not being followed by a portion of the population, and thus consumers are eating high-risk foods past their use-by dates.

Efforts to provide education regarding i) the meaning of date labelling terms; ii) the importance of shelf-life limitation for some products; iii) temperature control; iv) the availability and understanding of food storage guidance; as well as v) safe handling methods, could significantly reduce food waste and also improve food safety. In a world where hunger, malnutrition and food insecurity affect a significant portion of society, the appropriate level of protection for *L. monocytogenes* in foods must balance both food safety and the recognition of food waste that could be generated through precautionary recalls due to excess stringency in microbiological criteria.

It is important to remind consumers that product shelf-life is linked to storage conditions, and the risk assessment should address the impact of storage temperature on predicted illness and mortality. The Joint 2003 FDA/USDA *Listeria* Risk Assessment baseline model found that limiting the storage time for deli meat from the 28-day baseline to 14 days reduced the median number of cases of listeriosis in the elderly population from 228 to 197 (13.6 percent), and that shortening the storage time to 10 days further reduced the cases to 154 (32.5 percent) (FDA/FSIS, 2003). Modelling scenarios for other RTE foods (cantaloupe, polony) could reveal potential areas where public health improvements could be made. If cantaloupe was included as a product category in the updated risk assessment, product pathway exposure assessments could include *L. monocytogenes* contamination in the field upon harvest, during storage, cutting, distribution and from home sources that may further inform the risk of listeriosis to susceptible individuals.

Appropriate consumption

Individuals who may be susceptible to listeriosis due to age, pregnancy or immunosuppressive conditions should be advised to avoid high-risk foods and instead seek lower-risk alternatives.

In the case of frozen, non-RTE vegetables, adherence to package instructions for cooking at recommended temperatures prior to consumption is important, and consistency in label instructions may improve adherence to recommended guidance.

Consumer interest in high quality nutritious foods to support a healthy lifestyle has resulted in increased consumption of produce and potential concomitant exposure to *L. monocytogenes*. Fresh produce has emerged as an important source of foodborne illness outbreaks linked to *L. monocytogenes*, and this has become a global public health problem. Is there appropriate recognition by consumers of produce consumption being linked to outbreaks of listeriosis? Additional risk communication in this area may be necessary.

Many countries around the globe have managed to protect pregnant women from the risk of listeriosis. How do we get effective public health messaging to other parts of the globe and to other groups at risk? Why are older adults associated with increasing risk of listeriosis? Besides the fact that older adults are at a greater risk for hospitalization and death from foodborne listeriosis because organs and body systems go through changes as people age, another reason could be that some older consumers fail to adhere to use-by dates, may not have adequate refrigeration, and may leave opened RTE foods in the refrigerator for prolonged periods of time.

Data specific to older adult risk factors associated with listeriosis are insufficient (antacid use, proton pump inhibitors [PPI]) to communicate the risks of and prevent listeriosis in those concerned. Kvistholm Jensen *et al.* (2017) examined the use of PPI among individuals in Denmark and documented an increased risk of listeriosis. After adjusting for comorbidities and additional confounding variables, a 2.8-fold increased risk of acquiring listeriosis risk was associated with the use of PPI (Kvistholm Jensen *et al.*, 2017).

5.4   CONCLUSIONS

- Internationally, current regulations on *L. monocytogenes* contamination of RTE foods have either adopted a "zero tolerance" (non-detected in 25 g) approach for RTE foods or permit low levels in those foods that will not support the growth of the organism. Although finished product testing may be considered a control measure at the end of the production process, it gives only very limited information on the safety status of a food (Zwietering *et al.*, 2016).

- The majority of effort should be focused on management and control of this hazard in a more proactive way by implementing an effective food safety management system (FSMS) which includes environmental monitoring to assess incursions and persistence of the organism in processing facilities.

- Regulatory agencies should use a combination of finished product testing and environmental monitoring to verify proper implementation of *L. monocytogenes* control strategies by FBOs.

- *L. monocytogenes* isolates obtained through these processes should undergo subtyping (preferably WGS) with data retained in a database that includes subtyping data for human, food-processing environmental isolates, and if feasible, animal isolates.

- The approach varies between countries, with the CAs in some jurisdictions actively monitoring *L. monocytogenes* in RTE foods, while others evaluate company monitoring through an audit process.

- When an RTE food is found to exceed microbiological limits, a recall may be required. In addition, in some countries, when environmental monitoring confirms the presence of *L. monocytogenes* on food contact surfaces, this may also be a reason to initiate a recall.

- In terms of recalls, international networks such as the International Food Safety Authorities Network (INFOSAN), a global network of national food safety authorities, managed jointly by FAO and WHO, and the Rapid Alert System for Food and Feed from the European Commission (RASFF) are important networks that build capacity for the surveillance of foodborne diseases. The important links between INFOSAN and other regional, national and international networks are critical to improving collaboration across sectors and between programme to better manage food safety events such as recalls and outbreaks. Specific approaches regarding regulatory limits (e.g. "zero tolerance") might have an impact on the implementation of FSMS and particularly on the application of environmental monitoring programme.

- WGS of isolates assists with tracking and tracing of outbreaks, identifying virulence, and determining whether strains have established residence in food-processing facilities. However, in many countries, subtyping of isolates is not mandatory. In addition, a number of CAs do not have the resources and/ or technical knowledge to perform WGS.

- In order to reduce foodborne listeriosis, risk communication strategies must be developed to clearly communicate risk factors associated with product storage, shelf-life and appropriate consumption of RTE foods by vulnerable consumers.

- To reduce the risk, everyone has a role to play: FBOs, governments and consumers.
- Efforts to improve education regarding the meaning of date labels, the importance of shelf-life limitation for some products, temperature control, and the understanding of safe food handling practices could significantly improve food safety and lead to a reduction in food waste.
- Modelling scenarios for other RTE foods not previously covered in previous risk assessments (e.g. melons and deli meat) could reveal potential areas where public health improvements could be made.
- Fresh produce has emerged as an important source of foodborne illness outbreaks linked to *L. monocytogenes*, and this has become a global public health problem. However, risk-benefit considerations should be identified to highlight the relevant "trade-off" between health and potential risk.
- Additional risk communication in this area is recommended. This is increasingly important for produce that doesn't receive a kill-step, or for food products that may be consumed in their raw state, as might happen with frozen, non-RTE vegetables.
- Increasing age amplifies the likelihood of death in listeriosis cases. This reflects the fact that comorbidities and immunodeficiencies increase with age. Unfortunately, information on comorbidities is often lacking from epidemiological studies of listeriosis – more research into underlying causes is needed.

6

Laboratory methods for the detection and characterization of *Listeria monocytogenes* and *Listeria* spp.

6.1 INTRODUCTION

Listeria monocytogenes is an important foodborne pathogen that continues to cause foodborne outbreaks throughout the world. It is a major problem in the food industry because there are several factors that make this microorganism unique among foodborne pathogens as stated in Section 7.1. This dictates the need for control and prevention throughout the food chain and validated fit-for-purpose, qualitative and quantitative methods for the detection of *L. monocytogenes* and *Listeria* spp., both in foods and the environment. In fact, much of the growth in the global food microbiology testing market is due to a greater focus on environmental monitoring being driven by regulators, customers and auditors (Ferguson, 2020).

A review of current methods for the detection of *L. monocytogenes* and *Listeria* spp. focusing on validated methods for detection in food products and on environmental surfaces is presented in the following sections. The objective is to give an overview of current *L. monocytogenes* and *Listeria* spp. detection methods used by government and the food industry.

6.2 METHODS FOR THE DETECTION AND CHARACTERIZATION OF *L. MONOCYTOGENES* AND *LISTERIA* SPP. IN FOODS AND ENVIRONMENTAL SAMPLES

The food industry has been using microbiological methods for the detection of bacteria for decades, and many official/regulatory methods worldwide still rely on culture-based detection assays. These methods can be either qualitative, semi-quantitative or quantitative. Examples of the type of traditional methods that have been used to detect and/or enumerate *Listeria* spp. include cold enrichment, selective enrichment, direct plating and the most probable number (MPN).

Listeria spp. are generally non-fastidious and can be grown on many general media such as BHI broth/agar and TSA/TSB. The difficulties arise when they are present in foods that have a large background microflora and/or when the organisms become stressed due to physical or chemical inactivation methods, *e.g.* heat, HPP, antilisterial chemicals. It should also be noted that *L. monocytogenes* is not a competitor in the presence of other *Listeria* spp., especially *L. innocua*, and thus sometimes can be difficult to isolate from foods. In addition, viable-but non-culturable cells (VBNC) can occur naturally or under certain conditions as biofilm formation in foods and their environment.

The genus *Listeria* currently consists of 26 species. The only other pathogenic member besides *L. monocytogenes* is *L. ivanovii*, an organism that rarely infects humans but frequently causes listeriosis in ruminants. Together with *L. marthii*, *L. innocua*, *L. welshimeri*, *L. seeligeri*, *L. farberi*, *L. immobilis* and *L. cossartiae*, these two species form the "*Listeria sensu stricto*" group, one of the two distinct clades in the genus *Listeria*. All members of this clade (clade I) have been found in faeces or the GI tract of symptom-free animals and in foods of animal origin, suggesting a specific interaction of these species with mammalian hosts. Clade II, the "*Listeria sensu lato*" group, contains the species *L. fleischmanni*, *L. weihenstephanensis*, *L. rocourtiae*, *L. aquatica*, *L. cornellensis*, *L. riparia*, *L. floridensis*, *L. grandensis*, *L. grayi*, *L. newyorkensis*, *L. booriae*, *L. costaricensis*, *L. goaensis*, *L. thailandensis*, *L. valentina*, *L. portnoyi* and *L. rustica* which have been isolated from food-associated surfaces or the environment, animal farm environments, from mangrove swamps, soil samples, and agricultural water samples (Schardt *et al.*, 2017).

There are methods that will detect *Listeria sensu strictu* and not *Listeria sensu lato*, while others appear to detect both sensu strictu and *Listeria sensu lato*. As a result, questions are now arising as to whether a method should detect all *Listeria* species (that is both *Listeria sensu stricto* and *Listeria sensu lato*) to indicate the presence of pathogenic *Listeria* species, or if it is only necessary to detect *Listeria sensu strictu* species. Also, this would differ depending on whether the environment or final product is tested.

As such, many different *Listeria* enrichment broths and selective agars have been developed over the last 40 years to select and recover *Listeria* spp., each with their own unique formulations. Many cultural methods rely on the use of two different selective enrichments, followed by plating onto one or more selective media. Some of the selective agents that have been useful in the isolation of *Listeria* spp. include acriflavine, ceftazidime, lithium chloride, moxalactam, nalidixic acid, polymyxin B and potassium tellurite. In terms of selective agars, many have also been developed, both chromogenic and non-chromogenic.

There are many different factors that can influence the detection of *L. monocytogenes* in foods and environmental samples. Some of these include:
- the sensitivity, specificity, and detection limit of the method;
- the amount and number of samples analysed;
- the composition of the selective media and the temperature used for incubation;
- the number of different selective agars used (two is optimum, with one of them being a chromogenic agar);
- the number of colonies picked from agar plates for confirmation;
- the type of sample and food matrix;
- the microflora of the sample and food matrix;
- type of environmental surface sampled and the potential presence of biofilms;
- interference by other *Listeria* spp. (non-*monocytogenes*), especially *L. innocua*; and
- interference caused by the presence of multiple strains of *L. monocytogenes* in the same sample.

6.2.1   Cultural conventional methods

Conventional bacteriological methods are important because their use results in obtaining a pure culture of the organism which can be used for further studies. These methods remain the "gold standards" against which other methods are compared and validated. The disadvantages of these methods include the relatively long period of time that the protocols require for completion; "hands-on" manipulations; the requirement for many different chemicals, reagents, and media; the possibility of contaminating microorganisms; and the requirement for skilled analysts.

Many cultural methods use selective enrichment followed by plating methods to test for the detection or not detection of *L. monocytogenes* in a food sample (usually 25 g), with the typical detection limit being 1–5 CFU/test sample size. Enrichment is used to resuscitate any injured target organisms and increase their numbers while also diluting inhibiting compounds. Following enrichment, a selective medium is used to promote the growth of a target organism and decrease background flora, allowing for the isolation or detection of *Listeria*. Presumptive positives, if desired, can then be confirmed using additional physiological, biochemical, morphological, serological, and/or WGS testing. These cultural methods can take up to 4 days to obtain a presumptive positive or negative result, and confirmation of positive results can take up to a week (Gasanov, Hughes and Hansbro, 2005; Vălimaa *et al.*, 2015).

There are numerous official cultural-based methods for the detection of *L. monocytogenes* and *Listeria* spp., but many countries are using or have adapted the International Organization for Standardization (ISO) methods for quantitative and qualitative detection of *L. monocytogenes*. The reference methods for both the detection and enumeration of *L. monocytogenes* in food (Standards EN ISO 11290-1 & 2) have been internationally validated for five matrices, i.e. cold-smoked salmon, milk powdered infant formula, vegetables, cheese and the environment. Standard EN ISO 11290-1 is also considered a good method for the detection of *L. monocytogenes* in food-processing environments, especially for the food groups included in the study. For this method, the sensitivity rate varied from 91.1 percent to 100 percent, and the specificity rate varied from 97.6 percent to 100 percent (Gnanou-Besse *et al.*, 2019). Besides the European Union, some of the other countries that use the ISO methods for *L. monocytogenes* and *Listeria* spp. include India, Japan, Ireland, Australia (enumeration), New Zealand, Indonesia, South Africa and Brazil.

While most *L. monocytogenes* are detected following primary enrichment in half-Fraser broth, secondary enrichment in Fraser broth often results in more *L. monocytogenes* positive samples, especially when the original food sample contains low levels of *L. monocytogenes* or has a high level of competing microflora.

Expression of the results is often mentioned as "presence/absence of X in a test portion"; nevertheless, ISO-TC34/SC9 Microbiology agreed to change the expression of results in term of "detected or not detected in the test portion" to underline the limit of detection of the method not equal to 0 CFU in a test portion. Some of the important qualitative and quantitative methods that are being used for the detection of *Listeria* spp. and *L. monocytogenes* can be seen in Tables 5 and 6, respectively.

TABLE 5. A listing of some qualitative methods for the detection of *L. monocytogenes* (LM) and *Listeria* spp. including testing parameters

Method	Food source	Enrichment broth	Selective/ differential media	Total time	LM/*Listeria* spp. detection	Use with environmental sampling Y/N	Reference
ISO 11290-1:2017	All foods	HFB, FB	ALOA PALCAM or OXA	4–7 days	LM/*Listeria* spp.	Y	(International Organization for Standardization [ISO], 2017a)
FDA/BAM	All foods	BLEB	One of: OXA, PALCAM, MOX, LPM One of: R&F LMCPM, RAPID' L. mono, ALOA, OCLA, CHROMagar Listeria	4–7 days	LM	Y	(Hitchins, Jinneman, and Chen, 2017)
USDA-FSIS	Meat, poultry, egg, environmental	UMV	MOPS-BLEB MOX	4–7 days	LM	Y	(FSIS and USDA, 2021)
Health Canada MFLP-01	Meats, dairy products, fruit and vegetable products, smoked fish	UMV1, mFB	RAPID' L.Mono, PALCAM	4–7 days	LM	N	(Health Canada, 2015)
GB 4789.30 – 2016	All foods	LB1, LB2	PALCAM, CHROMagar Listeria	4–6 days	LM	N	(Ministry of Health of People's Republic of China, 2016)

* ALOA, Agar *Listeria* Ottaviani and Agosti Medium; BAM, Bacteriological Analytical Manual; BLEB, Buffered *Listeria* Enrichment Broth; FB, Fraser Broth; HFB, Half-Fraser Broth; FDA, Food and Drug Administration; ISO, International Organization for Standardization; LB, Listeria Broth; LPM, Lithium Chloride–Phenylethanol–Moxalactam Medium; mFB, modified Fraser Broth; MOPS-BLEB, Morpholine-Propanesulfonic Acid–Buffered *Listeria* Enrichment Broth; MOX, Modified Oxford *Listeria* Selective Agar; OCLA, Oxoid Chromogenic *Listeria* agar; OXA, Oxford Medium; PALCAM, PALCAM *Listeria* Selective Agar; R&F LMCPM, R&F *L. monocytogenes* Chromogenic Plating Medium; USDA-FSIS, U.S. Department of Agriculture—Food Safety and Inspection Service; UVM, Modified University of Vermont Broth.

CHAPTER 6 - LABORATORY METHODS FOR THE DETECTION AND CHARACTERIZATION OF *LISTERIA MONOCYTOGENES* AND *LISTERIA* SPP.

43

TABLE 6. A listing of some quantitative methods for the enumeration of *L. monocytogenes* (LM) and *Listeria* spp. including testing parameters.

Method	Food source	Enrichment broth	Selective/ differential media	Total time	LM/ *Listeria* spp. detection	Comments	Reference
ISO 11290-2:2017	All foods	BP	ALOA	3–5 days	LM/ *Listeria* spp.	Direct plating	(International Organization for Standardization [ISO], 2017b)
FDA/BAM	All foods	BLEB	One of: OXA, PALCAM, MOX, LPM One of: R&F LMCPM, RAPID' L. mono, ALOA, OCLA, CHROMagar Listeria	3–5 days	LM	Enumeration is performed using a combination of MPN and direct plating.	(Hitchins, Jinneman and Chen, 2017)
MFLP-74	All foods	BP	OXA and one of: ALOA, A.L. agar, BBL Chromagar Listeria, Brilliance Listeria Agar, LPM, MOX, PALCAM, Rapid L'Mono	1–2 days	LM	Direct plating	(Health Canada, 2015)

* ALOA, Agar *Listeria* Ottaviani and Agosti Medium; BAM, Bacteriological Analytical Manual; BLEB, Buffered *Listeria* Enrichment Broth; FB, Fraser Broth; HFB, Half-Fraser Broth; FDA, Food and Drug Administration; ISO, International Organization for Standardization; LB, *Listeria* Broth; LPM, Lithium Chloride–Phenylethanol–Moxalactam Medium; mFB, modified Fraser Broth; MOPS-BLEB, MorpholinePropane sulfonic Acid–Buffered *Listeria* Enrichment Broth; MOX, Modified Oxford *Listeria* Selective Agar; OCLA, Oxoid Chromogenic *Listeria* agar; OXA, Oxford Medium; PALCAM, PALCAM *Listeria* Selective Agar; R&F LMCPM, R&F *L. monocytogenes* Chromogenic Plating Medium; UVM, Modified University of Vermont Broth.

6.2.2 Environmental detection methods

It is common practice within the food industry to have an environmental monitoring programme for *Listeria* spp. Typically, sterile sponges with a neutralizing buffer are used to collect samples from food contact and non-food contact surfaces (Muhterem-Uyar *et al.*, 2015). The horizontal method ISO 18593 for environmental and surface sampling for microorganisms detection has been revised and is available (ISO 18593:2018). Environmental sampling for *Listeria* spp. is an important aspect of any *Listeria* food safety control programme and can be an early warning system for the presence of *L. monocytogenes* in the end product. The recurring presence of any *Listeria* spp. in the environment of a food facility signifies inadequate sanitation that can lead to the establishment of niches for *L. monocytogenes*. The combination of environmental monitoring with a rapid and aggressive response to positive test results allows food operators to continually improve their operations and reduce the potential of finding end products that are contaminated with *L. monocytogenes*. However, the success of these efforts relies heavily on the performance of the chosen detection method. Food companies are conducting both routine environmental sampling and more focused "seek and destroy" sampling programme. It is this latter strategy that is driving the increase in environmental testing that was mentioned earlier on.

6.2.3 Rapid and validated detection methods

The detection of *L. monocytogenes* in food samples by conventional culture methods is simple, sensitive, and inexpensive compared to molecular methods. However, conventional methods are laborious and time consuming as they require more than a week for detection and pathogen confirmation (Law *et al.*, 2015). Therefore, numerous rapid screening methods for *L. monocytogenes* in foods and the environment have been developed and are now in widespread use.

Desirable qualities in a rapid method include simplicity, cost effectiveness, a low limit of detection, online or real-time capabilities, and a low false-negative and false-positive rate. Rapid detection methods for *L. monocytogenes* and *Listeria* spp. often use a combination of steps, e.g. cultural enrichment followed by a rapid method for detection. Many of these tests are used as a rapid screen to identify any presumptively positive samples. All presumptively positive samples must be subjected to further confirmatory steps before they can be designated as positive for the presence of *Listeria* spp. (including *L. monocytogenes*).

CHAPTER 6 - LABORATORY METHODS FOR THE DETECTION AND CHARACTERIZATION OF *LISTERIA MONOCYTOGENES* AND *LISTERIA* SPP.

45

The most common rapid *Listeria* detection methods include antibody-based lateral flow or enzyme immunoassays (EIA) as well as polymerase chain reaction (PCR) based methods, e.g. isothermal or realtime PCR. Due to the recent advances in molecular technology, molecular methods have been used as alternatives to culture methods for food testing. Among these different technologies, PCR generally offers the greatest degree of sensitivity and specificity, even if some issues with PCR inhibitors may arise. The detection of *L. monocytogenes* is based on the detection of specific DNA or RNA sequences and such methods provide highly accurate and reliable results when compared to phenotypic methods. Nevertheless, molecular methods require specialized instruments and highly trained personnel (Law *et al.*, 2015).

Validated rapid methods are typically compared to a gold-standard reference method like EN ISO 11290, the FDA Bacteriological Analytical Manual (BAM) *Listeria* method, or the Health Canada MFLP-01 *Listeria* method. Methods for *L. monocytogenes* and *Listeria* spp. that have been validated by a governmental or independent certification body include Health Canada (https://www.canada.ca/en/health-canada/services/food-nutrition/research-programs-analytical-methods/analytical-methods/compendium-methods.html), Association Française de Normalisation (AFNOR) certification (https://nf-validation.afnor.org/en/food-industry/), Microval (https://microval.org/), NordVal (https://www.nmkl.org/index.php/en/22-nordval) and the AOAC (https://www.aoac.org/). In the European Union, official labs and the better commercial labs, in keeping with the requirements under Article 5.5 of Regulation 2073/2005, will use rapid methods which are validated by AFNOR certification to the ISO reference method (ISO 11290-1or-2 for *Listeria* spp. including *L. monocytogenes*), using the ISO 16140-2 protocol. Methods validated according to AFNOR are seen by labs as the gold standard in Europe, as they validate to the reference analytical (ISO) method stated in the regulation and follow the ISO 16140-2 protocol. The list of validated rapid methods according to ISO 16140-2 protocol is available (https://nf-validation.afnor.org/domaine-agroalimentaire/listeria-spp). In the United States of America, the AOAC is the American certification body that validates rapid methods. They use their own protocol to validate rapid methods against the FDA BAM method. Some of the rapid tests available on the European Union market will only have AOAC certification, while some will have both, knowing that it is essential for their commercial success in both the United States of America and European Union markets. In Canada, companies need to use methods that are listed in the Compendium of Methods for regulatory/"official" samples (https://www.canada.

ca/en/health-canada/services/food-nutrition/research-programs-analytical-methods/analytical-methods/compendium-methods.html).

6.2.4 Innovative and emerging detection methods

On the research front, there are many emerging methods for the detection of *L. monocytogenes* and *Listeria* spp. including fluorescence (Guo *et al.*, 2020; Li *et al.*, 2020; Meile *et al.*, 2020), single primer isothermal amplification (Yang et al., 2020), electrochemical biosensing (Chen *et al.*, 2020; Silva *et al.*, 2020), and a duplex lateral flow dipstick (DLFD) test combined with loop-mediated isothermal amplification (LAMP) (Ledlod *et al.*, 2020; Wang *et al.*, 2020). In addition, WGS approach through metagenomics by direct sequencing of microbiological enrichments should be a powerful promising method in the future, leading at the same time to the detection and characterization of *L. monocytogenes* within 2 days (Ottesen *et al.*, 2020). Although not specific to *Listeria* spp., some newer rapid phage-based methods also show promise in terms of being able to rapidly detect viable pathogens in food (Foddai and Grant, 2020). However, none of the above methods have been validated for the detection of *Listeria* spp. in foods.

Historically, the foundation of public health surveillance for enteric bacteria has been a reliance on accurate identification and recovery of causative agents by stool culture (Imdad *et al.*, 2018). However, culture-independent diagnostic tests (CIDTs) are increasingly being used in hospitals and clinical labs to detect foodborne pathogens in clinical samples such as stool, food and cerebrospinal fluid.

The advantages of CIDT include i) sensitivity (80 percent to 100 percent) and specificity (88 percent to 99 percent); ii) more rapid results compared to traditional culturing methods, that is CIDTs can identify the general type of bacteria causing illness within hours, without having to culture or grow the bacteria in a lab; and iii) the ability to detect multiple pathogens in one test. The disadvantages of CIDT include i) unless reflex culturing is done, CIDT-identified bacteria are not available to perform advanced genetic characterization that is used to identify clusters, outbreaks, emerging infections or perform antimicrobial resistance (AMR) testing; this could be problematic in the future for early detection and tracing of outbreaks; ii) false-positives/negatives; iii) relatively high costs; and iv) when isolates are not available from the clinical laboratory, the burden of isolate recovery is transferred to under-resourced public health labs (Imdad *et al.*, 2018).

CHAPTER 6 - LABORATORY METHODS FOR THE DETECTION AND CHARACTERIZATION OF *LISTERIA MONOCYTOGENES* AND *LISTERIA* SPP.

47

Pathogen recovery from CIDT-positive specimens will vary depending on the pathogen type. In the clinical area, CIDT is a fairly normal practice now. PCR is mainly being done, and if positive, the stools are sent to reference labs who do WGS and culture the pathogens when necessary. Methodology advances in CIDT for diagnostics and public health are still in the early stages. High multiplex PCR and shotgun metagenomics of clinical, food and environmental samples are potential candidates to replace WGS of pure cultures. Shotgun metagenomics to detect common enteric bacterial pathogens in stool is nearly as fast and accurate as culture, with a similar cost to some syndromic PCR panels. However, PCR testing is still faster, more sensitive, and cheaper and will continue to be the method of choice for routine testing for the immediate future.

6.2.5 Typing methods

It is important to subtype pathogenic strains of *L. monocytogenes*, e.g. during foodborne illness outbreaks, for trend analysis, as well as for "seek and destroy" operations in food establishments. This has been (and still is in some countries) conventionally performed by methods such as ribotyping, MALDI-TOF, spectra profiling, pulsed-field gel electrophoresis (PFGE), multi-locus sequence typing (MLST), multi-locus variable number tandem repeats analysis (MLVA) and multi-virulence-locus sequence typing (MVLST). Worldwide, these methods are being or have been replaced by the WGS approach, which can simultaneously provide information on serotype, antimicrobial resistant (AMR) genes and mobile genetic element, virulence markers, and also allow clustering based on single nucleotide polymorphism (SNP) analysis and cgMLST or wgMLST complex typing (CT type) using a unique assay (Moura *et al.*, 2016; Félix *et al.*, 2018; Brown *et al.*, 2019; Nouws *et al.*, 2020; Pietzka *et al.*, 2019; Ronholm *et al.*, 2016). Therefore, WGS has been fast replacing PFGE and other molecular methods because it has the highest discriminatory power. By sequencing the entire target bacterial genome which yields an incomparable depth of data, WGS offers the capacity to differentiate the phylogenic association better than other advanced methods (Ribot *et al.*, 2019).

The discriminatory power of these molecular typing methods is summarized in Figure 4.

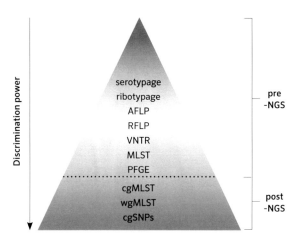

Amplified fragment-length polymorphism (AFLP), Restriction fragment-length polymorphism (RFLP), Variable number tandem repeat (VNTR), Multi-locus sequence typing (MLST), Pulsed-fiend gel electrophoresis (PFGE), core genome MLST (cgMLST), whole genome MLST (wgMLST), core genome SNPs (cgSNPs)

FIGURE 4. Discrimination power of typing methods pre- and post-whole genome sequencing (WGS).

Source: Reproduced with permission from Radomski, 2020.

The complexity of outbreaks was highlighted by the findings associated with the 2011 United States of America cantaloupe outbreak which identified a total of five outbreak-associated subtypes of *Listeria* (McCollum *et al.*, 2013). WGS was used to assess the genome level diversity of *L. monocytogenes* strains during the investigation of the 2010–2015 United States of America ice cream outbreak (Chen *et al.*, 2017). WGS (SNP analysis) differentiated outbreak-associated *L. monocytogenes* strains from epidemiologically unrelated strains that matched outbreak PFGE/MLST profiles. SNP analysis allowed simultaneous identification of a clonal complex (CC5) and discrimination of different outbreak strains in the same clone. This scientific evidence demonstrated the necessity of molecular tools for effectively linking outbreak strains with resulting infected human cases, and for determination of the pathways of food contamination (Farber *et al.*, 2021).

CHAPTER 6 - LABORATORY METHODS FOR THE DETECTION AND CHARACTERIZATION OF *LISTERIA MONOCYTOGENES* AND *LISTERIA* SPP.

49

In the same way, a multicountry outbreak of 47 listeriosis cases in Austria, Denmark, Finland, Sweden and the United Kingdom caused by *L. monocytogenes* PCR serotype 4b sequence type (ST6) was investigated through WGS analysis, with frozen corn and possible other frozen vegetables recognized as vehicles of infection (Felix *et al.*, 2018).

Although the lack of standardization of next-generation sequencing (NGS) methods and interpretation has been noticed, a standardization effort is underway through the ISO/DIS 23418 project entitled "Whole genome sequencing for typing and genomic characterization of foodborne bacteria - General requirements and guidance", which is being conducted in the framework of ISO/TC34/SC9 (Microbiology of the Food Chain). Furthermore, a cgMLST method has been validated internationally by different public health laboratories (Moura *et al.*, 2016) and an international platform for MLST and cgMLST curation, deposit and analysis established (BIGSdb Listeria; https://bigsdb.pasteur.fr/listeria/listeria.html).

6.3 CONCLUSIONS

- Most countries outside of North America appear to rely on ISO methods both for the detection and enumeration of *Listeria* spp. and *L. monocytogenes.*
- Several recognized bodies have validated methods for *Listeria* spp. and *L. monocytogenes* that can be used and are fit-for-purpose. However, not all of them have been validated for all food groups.
- The major organizations involved in some aspects of alternative method recognition and/or validation include i) ISO, ii) AOAC, iii) AFNOR, iv) NordVal, v) Microval, vi) FDA/USDA, and vii) Health Canada.
- Among the numerous commercial methods, each one has their own unique advantages and disadvantages. Furthermore, each should be considered as a first approach for rapid sample screening giving a presumptive result, which can be especially valuable during an outbreak investigation. Any positive result would require a confirmation step using an official validated method from a recognized body.
- WGS has a high level of discrimination at the nucleotide level and provides all the information (virulence markers, AMR or biocide resistance genes, cgMLST) for *L. monocytogenes* strains. It significantly enhances outbreak investigations and contamination source attribution.
- WGS is becoming less expensive and more available and is being more frequently used by public health and food safety labs in upper-middle-income economies. However, bioinformatics, for example, calculation and storage of

data, standardization of protocols, pipeline development, and harmonization of data nomenclature is needed.

- CIDT testing is becoming more popular for analysing clinical samples for the presence of *L. monocytogenes*; however, it is essential to recover the pure culture of the clinical strain for further characterization, as well as to confirm obtained results with clinical symptoms of the patient to avoid false-positive results.

- Metagenomics (direct sequencing from microbiological selective enrichments) appears to be a promising method in the future, leading at the same time to the detection and characterization of *L. monocytogenes* within 2 days.

- Strengthening surveillance in individual countries by harmonizing microbiological methods and providing epidemiologic tools for investigations will be a key step in reducing the public health burden of listeriosis, even as the population at-risk grows (Hedberg, 2006).

6.4 RECOMMENDATIONS

- Standardized and validated culture methods for isolation (e.g. ISO, FDA, USDA, Health Canada) should be used to obtain pure isolates for confirmation and further characterization, including antibiotic sensitivity, biocide tolerance and subtyping, preferably using WGS-based cgMLST and/or SNP analysis.

- The use of conventional methods should be complemented by emerging methods such as qPCR, and MALDI TOF MS for rapid confirmation, particularly in outbreak situations or for clinical detection.

- There are only a handful of recognized validated enumeration methods for low levels of *L. monocytogenes* in foods; more research needs to be done in this area, especially for dose-response assessment and hazard characterization.

- It is recommended that at least two different plating media, based on different principles, be used, with one of them being a chromogenic medium such as the "Agar Listeria according to Ottaviani and Agosti" (ALOA) to distinguish *L. monocytogenes* from other *Listeria* species.

- Methods for the detection of viable but nonculturable (VBNC) bacteria, stressed cells and those in biofilms should be developed, especially for environmental samples. These methods should distinguish between live, dead or sublethally injured cells.

7

Hazard characterization

7.1 INTRODUCTION

Listeria monocytogenes, the causative agent of listeriosis, is an important foodborne pathogen that continues to cause global foodborne outbreaks. In fact, an increasing number of outbreaks have been observed internationally and are linked to foods not usually recognized as causing foodborne listeriosis (Desai *et al.*, 2019). However, it is still considered a rare disease, with an annual incidence in high-income countries of between two to five cases per 1 000 000 persons. A definition of invasive listeriosis can be found in Section 2.3.2.

There are several factors that make this microorganism unique among foodborne pathogens.

These include that it:
- is very widespread in the environment;
- can grow at refrigeration temperatures;
- can survive in the environment of food-processing plants for months/years;
- mainly affects at-risk individuals;
- is associated with a high case-fatality rate of around 15 to 30 percent and a high hospitalization rate (> 90 percent); and
- has many virulence factors that can help to promote its intracellular lifestyle.

7.2   SEVERITY OF LISTERIOSIS

Invasive listeriosis occurs when *L. monocytogenes* is ingested and the organism survives its passage through the GI tract and then attaches and enters a variety of intestinal cells leading to the invasion of otherwise sterile body sites. Foodborne listeriosis infections cause a wide spectrum of illness, ranging from febrile gastroenteritis (non-invasive listeriosis) to invasive clinical forms, including septicemia, meningitis and meningoencephalitis. During pregnancy, listeriosis can cause miscarriage, premature birth, severe illness in a newborn child or stillbirth in rare cases. The organs most often infected are the pregnant uterus, liver, spleen and brain. Maertens de Noordhout *et al.* (2014) gathered data on human listeriosis through a systematic review of peer-reviewed and grey literature published from 1990–2012. Among all listeriosis cases, 20.7 percent and 79.3 percent were perinatal and non-perinatal infections, respectively. The overall case-fatality rates for perinatal and non-perinatal cases was 14.9 percent and 25.9 percent, respectively. Perinatal listeriosis can be defined by the evidence of *L. monocytogenes* in any sample of maternal, foetal or neonatal origin. It is a severe condition associated with foetal loss, or early or late neonatal infection with frequent sepsis or meningitis (Charlier *et al.*, 2022). In the large MONALISA national prospective observational cohort study performed in France, of the 818 cases enrolled in the study, there were 107 (13.1 percent) maternal–neonatal infections, 427 (52 percent) cases of bacteraemia and 252 (30.8 percent) cases of neurolisteriosis (Charlier *et al.*, 2017).

Most cases of listeriosis appear to be sporadic, although a portion of these sporadic cases may be unrecognized common-source clusters. However, with the advent of WGS, we are seeing more recognized national and cross-border clusters of listeriosis, as well as detecting smaller outbreaks. The trend has been to recognize fresh produce as an increasing vehicle of foodborne listeriosis, although outbreaks linked to meat (Thomas *et al.*, 2020; Halbedel *et al.*, 2020), dairy (Amato *et al.*, 2017; Jackson *et al.*, 2018) and fish (Schjørring *et al.*, 2017) products continue to occur.

Listeria monocytogenes is known to cause acute, self-limited, febrile gastroenteritis in healthy persons and has been the incriminating microorganism in at least seven outbreaks of foodborne gastroenteritis. These outbreaks have generally involved the ingestion of high doses ($> 10^6$ CFU/g) of *L. monocytogenes* by otherwise healthy individuals. In large outbreaks where the majority of cases present with diarrhea and the typical GI pathogens are not detected, *L. monocytogenes* should be considered a causative agent.

Nevertheless, in this report, we only include invasive cases of listeriosis because this is where most of the public health burden due to this microorganism lies. In addition, several laboratory criteria for the diagnosis of listeriosis such as in the United States of America or Europe include the isolation of *L. monocytogenes* from a normally sterile site, for example, blood or cerebrospinal fluid or, less commonly, joint, pleural or pericardial fluid.

7.3 GENERAL RISK FACTORS FOR FOODBORNE LISTERIOSIS

There are several risk factors for foodborne listeriosis (see Section 4.2). These include the presence and growth potential of *L. monocytogenes* in the food, the amount and type of *L. monocytogenes* ingested and the host susceptibility. The at-risk populations for listeriosis include pregnant women, neonates, the elderly and immunocompromised individuals, for example, cancer patients, AIDS patients, people with liver problems and so on (Goulet *et al.*, 2012a). It should be recognized, however, that healthy individuals can become ill with a severe form of listeriosis. For example, in the caramel apple listeriosis outbreak, three invasive illnesses (meningitis) occurred among otherwise healthy children aged 5–15 years (Angelo *et al.*, 2017).

7.4 PATHOGENESIS OF *L. MONOCYTOGENES*

Listeria monocytogenes has an abundance of virulence factors, a unique intracellular lifestyle, and an ability to cross multiple host barriers, for example, the placenta and blood-brain barrier. For these reasons, it is also used as a model bacterium to study how pathogenic intracellular bacteria mediate host-pathogen interactions.

Once in the GI tract, *L. monocytogenes* makes contact with the intestinal epithelium, translocates across the intestinal epithelial barrier into the lamina propria and then disseminates via the lymph and blood to the liver and spleen. Once the organism gains entry into the liver, it can translocate into the gallbladder through bile canaliculi, where it can multiply extracellularly. This allows for the re-entry of *L. monocytogenes* cells into the GI tract through the biliary ducts. The organism can also potentially cross the blood-brain barrier or the placental wall (Pizarro-Cerdá *et al.*, 2012; Radoshevich and Cossart, 2018; Lecuit, 2020).

All *L. monocytogenes* strains are considered to have the potential to cause illness in humans and be a health risk, especially in susceptible individuals. The relative potential of a single strain of *L. monocytogenes* to cause severe disease in humans can be assessed based on virulence gene content and sequences (e.g. the presence of premature stop codons in virulence genes) and the at-risk status of the individual consuming the contaminated food.

7.5 VIRULENCE GENES AND FACTORS INVOLVED IN ATTACHMENT, ENTRY, MULTIPLICATION AND CELL-TO-CELL SPREAD

The following brief description of the pathogenesis of *L. monocytogenes* is intended to provide an overview of potential genes where gene modifications (deletions, premature stop codons, etc.) can be used to distinguish between strains of variable virulence. Once inside the gut, *L. monocytogenes* can invade intestinal epithelial cells with the help of a full-length internalin A (InlA) through either the Peyer's patches, intestinal villi or goblet cells (Jacquet *et al.*, 2004; Drolia and Bhunia, 2019); invasion of Peyer's patches can occur independent of an intact InlA. In goblet cells, *L. monocytogenes* can transcytose across the cell within a vacuole, and in some macrophages, it can multiply in spacious *Listeria*-containing phagosomes (SLAPs) (Radoshevich and Cossart, 2018). Surface located InlA and inlB bind to epithelial cells (E-cadherin surface receptor) and hepatocytes (c-Met surface receptor), respectively.

After cell uptake, the bacterium can lyse the vacuole through the action of a hemolysin, Listeriolysin O (LLO), and two phospholipases, PlcA and PlcB. Once inside the cytosol, *L. monocytogenes* starts expressing the hexosephosphate transporter (Hpt), which enables it to make use of the sugar sources available inside the mammalian cell, for example, glucose 6-phosphate, thus enabling bacterial growth. Through the action of another virulence factor, ActA, the Arp2/3 complex is recruited, which results in facilitating actin polymerization and protrusion of the bacterium to the adjacent cell. Once there, the double membrane vacuole is lysed, again by the action of LLO, PlcA and PlcB (Hamon, Bierne and Cossart, 2006; Pizarro-Cerdá *et al.*, 2012; Radoshevich and Cossart, 2018). Internalin C, another internalin in the family of about 25 internalins, appears to play a role in the process by allowing bacteria to more efficiently form bacteria-filled cell protrusions which promote bacterial spreading within infected host tissue, while internalin P (inlP), conserved in virulent *L. monocytogenes* and absent in non-pathogenic listeriae, appears to be critical for placental infection (Faralla *et al.*, 2016, 2018). ActA and LLO also appear to be involved in *L. monocytogenes* replication in the placenta and dissemination into foetal tissues (Charlier, Disson and Lecuit, 2020).

7.6 MASTER VIRULENCE REGULATOR AND KEY TRANSCRIPTIONAL REGULATORS

After host cell adhesion and uptake, the transcriptional regulator PrfA, which is a member of the Crp/Fnr family of regulators, as well as SigB, play a major role in the whole lifestyle of intracellular growth and spread (de las Heras *et al.*, 2011). SigB is the regulator of general stress response, virulence and resilience (Liu *et al.*, 2019). In fact, the virulence regulation of *L. monocytogenes* greatly depends on PrfA, which controls the expression of a broad array of genes, including major virulence factors, and is thus regarded as the *L. monocytogenes* master virulence regulator (de las Heras *et al.*, 2011). It acts by binding as a dimer to a palindromic consensus binding site, known as the PrfA-box, lying upstream (~40 bp) of the transcriptional start sites of genes that are positively controlled by PrfA. The absence of a non-functional form of PrfA severely attenuates the virulence of *L. monocytogenes* in cell culture infection models. Its expression and activity appear to be controlled at several levels, i.e. transcriptional, post-transcriptional and post-translational (Radoshevich and Cossart, 2018).

Outside the host, the expression of PrfA-regulated genes is low, but when entering a host, PrfA becomes activated and turns on the expression of PrfA-regulated virulence genes. For activation, PrfA requires binding of the cofactor glutathione. Glutathione binding stabilizes the DNA-binding helix-turn-helix (HTH) motif in a conformation compatible with DNA binding, thereby allowing transcription of PrfA-regulated virulence genes (Tiensuu *et al.*, 2019; Hansen *et al.,* 2020). In addition to PrfA, another key transcriptional regulator appears to be sigma factor B (sigB), and some type of regulatory co-operation between the two appears to be required for responding to environmental stress conditions and for host infection (Oliver *et al.*, 2010; Gaballa *et al.*, 2019). Many other regulators such as VirR (Mandin *et al.*, 2005), Hfq (Christiansen *et al.*, 2004) and MogR (Shen and Higgins, 2006) also appear to contribute to a lesser extent to the *L. monocytogenes* virulence regulatory network.

7.7 ESSENTIAL VIRULENCE FACTORS OF *L. MONOCYTOGENES* AND THEIR POTENTIAL ROLE IN INFORMING RISK MANAGERS

The ability of *L. monocytogenes* to attach, enter, multiply, spread and invade various human tissues and organs depends on numerous virulence factors. Current evidence points to the most important, these being encoded in the *inlA-inlB* locus and in the pathogenicity islands LIPI-1, LIPI-3 and LIPI-4. It should be noted that expression of both the *inlA-inlB* locus and LIPI-1 is regulated by the transcriptional regulator PrfA (Quereda *et al.*, 2019).

7.7.1 *Listeria* pathogenicity island I (LIPI-1)

Pathogenicity island LIPI-1 is conserved among all pathogenic *L. monocytogenes* strains and contains key genes including *prfA, plcA, hly, mpl, actA*, and *plcB*, that participate in host invasion and cellular proliferation (Vazquez-Boland *et al.*, 2001), all of which are controlled by the main virulence activator PrfA. LIPI-1 is present with an identical structure in both *L. monocytogenes* and *L. ivanovii*, but is not present in *L. innocua* (except for hemolytic *L. innocua*) and *L. welshimeri*.

7.7.2 *Listeria* pathogenicity island I (LIPI-3)

Listeria monocytogenes pathogenicity island 3 (LIPI-3) consists of eight genes, which are linked to the production of listeriolysin S (LLS) and hypervirulence. LLS is a post-translationally modified hemolysin that has been shown to play a role in the survival of *L. monocytogenes* in polymorphonuclear leukocytes (PMNs) and also to contribute to the virulence of the organism in the murine model (Cotter *et al.*, 2008). However, Quereda *et al.* (2016, 2017) found that LLS does not contribute to virulence *in vivo* once the intestinal barrier has been crossed. Instead, it acts as a bacteriocin, which disrupts the intestinal microbiota, thus enhancing the colonization of *L. monocytogenes*. Since the *llsX* gene may be specific for *L. monocytogenes*, it can be used as a marker for the LIPI-3 pathogenicity island. Although LIPI-3 appears to be found only in lineage I isolates of *L. monocytogenes*, a corresponding gene cluster or its remnants has been found in a number of *L. innocua* strains (Clayton *et al.*, 2014).

7.7.3 *Listeria* pathogenicity island 4 (LIPI-4)

Maury *et al.* (2016) reported a newly identified pathogenicity cluster, known as *Listeria* pathogenicity island 4 (LIPI-4), which mainly includes the cellobiose-family phosphotransferase system (PTS) and has also been linked to hypervirulence. The operon contains a cluster of six genes encoding a sugar transport system involved in neural and placental infection. This clonal complex (CC) 4-associated PTS is specifically involved in the selective tropism of *L. monocytogenes* for the central nervous system (CNS) and the placenta.

A listing of all the pathogenicity islands, along with their major virulence genes and association with hypo- or hyper-virulence can be seen in Table 7.

TABLE 7. Characteristics of the pathogenicity islands of *L. monocytogenes* and *L. ivanovii*

Pathogenicity island	Main virulence gene (others)	Associated with hypervirulence	Comments
LIPI-1	*prfA, hly, plcA/B, mpl, actA*	Yes	Widely distributed in Lm; controlled by PrfA
LIPI-2	Encodes sphingo-myelinase and numerous internalins	No	Specific for *L. ivanovii*
LIPI-3	Listeriolysin S (LLS); *llsX, llsA, llsB, llsD, llsG, llsH, llsP, llsY*	Yes	Only present in a subset of lineage I strains and some atypical *L. innocua*; bacteriocin that can alter the host intestinal microbiota and promote intestinal colonization
LIPI-4	Mainly includes the cellobiose-family phospho-transferase system (PTS-*ptsA*); *ptsA, licC, licB, licA, glvA*	Yes	Only found in some lineage I strains; also found in *L. innocua* strains; associated with neurolisteriosis

7.8 GROUPINGS OF *L. MONOCYTOGENES* THAT COULD BE IMPORTANT FOR RISK MANAGERS

As *L. monocytogenes* is a highly heterogenous species, there are numerous classification schemes for the organism. In general, it can be divided into 14 serogroups, four PCR serogroups and four evolutionary lineages (Table 8). Further subdivisions include sequence types (ST), clonal complexes (CC), both based on MLST typing, as well as sublineages (SL) and CT types based on cg-MLST methods using WGS.

TABLE 8. General classification of *L. monocytogenes* by various methods

Lineage	Serotype[a]	Examples of clonal complex (CC) types	Example of sequence types (ST)	Comments
I	1/2b, 3b, 4b, 4b(V), 4d, 4e	1[b], 2[b], 3, 4[b], 6[b] (1 and 4 most strongly associated with MN and CNS infections)	1, 2	Over-represented among human isolates; CC1 – ruminant infection
II	1/2a, 1/2c, 3a, 3c, 4h[d]	9[c], 121[c]	8, 9, 87, 121, 155	Over-represented among food/natural envs.
III	4a, 4ab, 4b (atypical), 4c			Most isolates from ruminants
IV	4a, 4ab, 4b (atypical), 4c			Rare; most isolates from ruminants

[a] Doumith *et al.* (2004) established a PCR molecular serogrouping method that subdivides *L. monocytogenes* into four groups: I.1 (1/2a-3a), II.2 (1/2b-3b-7), I.2 (1/2c-3c), II.1 (4b-4d-4e) and III (4a-4c). An additional PCR serogrouping profile, PCR IVb variant 1 (IVb-v1), was discovered by Leclercq *et al.* (2011).
[b] Hypervirulent association.
[c] Hypovirulent association.
[d] Serotype 4h, HSL-II isolates are highly virulent and exhibit higher organ colonization capacities than well-characterized hypervirulent strains of *L. monocytogenes* in an orogastric mouse infection model. The isolates contain the *L. monocytogenes* pathogenicity island (LIPI)-1, a truncated LIPI-2 locus, and a unique cell wall teichoic acid structure (Yin *et al.*, 2019).

7.9 SEROTYPING

When *L. monocytogenes* first emerged as an important foodborne pathogen as a result of an outbreak in Canada linked to the consumption of contaminated coleslaw, serotype information was seen to be an important factor in the hazard characterization of the organism. Although there are currently 14 serotypes of *L. monocytogenes*, over 90 percent of listeriosis cases are associated with three serotypes, namely 1/2a, 1/2b, and 4b, with serotype 4b causing roughly over half of these cases. Serotyping of the organism has gradually fallen out of favour because i) serotype in itself is not a virulence factor, ii) the sera used to determine the serotype are less frequently being produced by a few commercial companies so they are hard to obtain and expensive, and iii) serotype is not routinely used in outbreak investigations due to its lack of discriminatory power. Some potential exceptions include i) the recent observation that a rare variant of serotype 4b, termed Ivb-v1 (or 4bV), has been the cause of several recent outbreaks and may be undergoing expansion (Leclercq *et al.*, 2011; Burall *et al.*, 2017); and ii) a new serotype (4h), belonging to the hybrid sublineage of the major lineage II (HSL-II) which appears hypervirulent and which in an orogastric mouse infection model exhibits greater organ colonization capacities than well-characterized hypervirulent strains of *L. monocytogenes* (Yin *et al.*, 2019); and iii) molecular serotyping being used as a rapid system for separating the four major *L. monocytogenes* serovars most frequently isolated from food and patients (1/2a, 1/2b, 1/2c, and 4b) into distinct groups during an investigation, or for the management of critical control point for FBOs (Doumith *et al.*, 2004). However, nothwithstanding the above, serotype information has limited usefulness for epidemiological surveillance, and serotype data alone is not sufficient for assessing the health risk of individual strains of *L. monocytogenes*.

7.10 LINEAGES

With regards to lineages, *L. monocytogenes* has a clonal population structure that is organized into four phylogenetic lineages with lineages I and II being the major lineages causing human illness (Nightingale *et al.*, 2005; Maury *et al.*, 2016). Differences in virulence can be observed in specific genomic lineages (Table 8). Historically, lineage I strains have been responsible for the majority of severe disease and large outbreaks, while lineage II strains have been observed mostly as sporadic clinical cases. While this is still the case in some regions, some countries are observing a trend indicating the emergence of lineage II as a significant cause of clinical disease cases and outbreak events (Orsi *et al.*, 2011; Haase *et al.*, 2014). While hypervirulent and hypovirulent strains of *L. monocytogenes* appear to predominantly belong to lineage I and II, respectively (Jacquet *et al.*, 2004; Orsi *et al.*, 2011; Maury *et al.*, 2016), lineage I does include some hypovirulent strains characterized by *inlA* premature stop codons (PMSCs).

7.11 MODIFIED VIRULENCE GENES

The presence of PMSCs in the internalin A gene (*inlA*) in many serotype 1/2c isolates of *L. monocytogenes* has been hypothesized to be responsible for the decreased incidence of listeriosis cases attributable to this serotype (Nightingale *et al.*, 2008; Jacquet *et al.*, 2004; Tamburro *et al.*, 2010). It should be noted that up to 18 different PMSCs exist in *inlA* (Maury *et al.*, 2016; EFSA, 2018) and are common among food-related lineage II strains, while they are rare among lineage I strains (Ducey *et al.*, 2007; Van Stelten *et al.*, 2010; Maury *et al.*, 2016).

Epidemiological data suggests that isolates with *inlA* PMSCs are > 3 log less likely to cause disease, as compared to strains that encode the full length InlA (Chen *et al.*, 2011). This correlates with what has been observed in the past with food and clinical isolates of *L. monocytogenes* serotype 1/2c, that is, they have historically been found frequently in some foods, but rarely in human cases (Doumith *et al.*, 2004; Tamburro *et al.*, 2010; Orsi *et al.*, 2011). There can also be modifications in the internalin B gene (*inlB*). Both phylogenetically defined isoforms (Chalenko *et al.*, 2019), deletions of greater than 100 nucleotides (Kurpas *et al.*, 2020), as well as point mutations that result in nonsense mutations, have been reported (Quereda *et al.*, 2019). With regards to the latter, the strain that caused the 1985 soft-cheese California listeriosis outbreak did have a point mutation in the inlB, a mutation that apparently made the strain less virulent (Quereda *et al.*, 2019). While naturally occurring virulence attenuating mutations have also been reported in other genes (e.g., *prfA, mpl*), these mutations appear to be rather infrequent and hence appear to be less relevant from a public health perspective (Velge *et al.*, 2007; Roche *et al.*, 2009; Maury *et al.*, 2017).

7.12 SEQUENCE TYPES

With regard to sequence types (STs), many different STs have been the cause of foodborne listeriosis (Chenal-Francisque *et al.*, 2011), but some have predominated, e.g. ST6 has been the cause of a number of outbreaks reported around the world (Table 9). Apparently unique hypervirulent sequence types exist in various parts of the world, such as has been described for ST 87 in China (Li *et al.*, 2020) and Spain (Pérez-Trallero *et al.*, 2014).

TABLE 9. Serotype, ST and CC of strains that have caused some recent outbreaks of foodborne listeriosis

Country/food	Serotype	Clonal complex	Sequence type	Year	Reference
South Africa/ polony	4b	-	6	2017/18	Smith *et al.*, 2019
Germany/ sausage		-	6	2018/19	Halbedel *et al.*, 2020
Canada/deli meat	1/2a	8	120, 292	2008	Currie *et al.*, 2015
Canada/coleslaw	4b	1	1	1981	Schlech *et al.*, 1983

(cont.)

USA/stone fruit	4b	-	382	2014	Chen *et al.*, 2016
USA/mung bean sprout	4b	-	554	2014	Garner and Kathariou, 2016
European Union/CS fish		8	-	2014/2019	EFSA and ECDC, 2019a
European Union/RTE meats European Union/frozen corn	4b 4b	- -	6 6	2017/19	EFSA and ECDC, 2019b EFSA and ECDC, 2018b
Spain/RTE meat	4b	388	388	2019	M. Medina, personal communication, 2020
Switzerland/meat pate	4b		6	2016	Althaus *et al.*, 2017
USA/hot dogs	4b	6	-	1998/99	CDC, 1998
USA/mushrooms		7	7	2019	CDC, personal communication, 2020
USA/eggs		155	372	2019	CDC, personal communication, 2020
USA/deli meat/cheese		321	2041	2019	CDC, personal communication, 2020
Denmark/smoked fish (2 outbreaks)	1/2a 4b	- -	391 6	2013/15 2013/15	Gillesberg Lassen *et al.*, 2016

7.13   CLONAL COMPLEXES

A clonal complex or CC is defined by a 7-locus MLST scheme and consists of a group of strains whose STs differ by no more than one allele from at least one other ST in the group. MLST-based classification into CCs allows for definition of subgroups within lineages. Chenal-Francisque *et al.* (2015) genotyped 300 isolates from 42 countries on five different continents and found that the strains represented 111 STs which grouped into 17 CCs, showing the existence of a few prevalent and globally distributed clones. In general, there appear to be three distinct patterns among all the major *L. monocytogenes* clones (Maury *et al.*, 2019):

- clones that are host associated, highly prevalent in dairy products, exhibiting a low adaptation to food production environments and rarely harbouring benzalkonium chloride tolerance genes (i.e. CC1 and CC4);

- clones with low adaptation to the host, but that persist efficiently in food-production environment, possibly aided by reduced sensitivity to disinfectants (biocides) due to a high prevalence of benzalkonium chloride tolerance genes (i.e. CC9 and CC121). The co-selection of tolerance to disinfectants (e.g. quaternary ammonium) used in the food-processing industry and resistance or altered susceptibility to some antibiotics (e.g. ciprofloxacin, gentamicin, amoxicillin) upon exposure of *L. monocytogenes* to antimicrobial agents may lead, inadvertently, to emerging and underestimated issues in food safety and public health with a concomitant increased risk of a poor disease outcome; and
- intermediary clones that may be in the process of transitioning from host-associated to saprophytic lifestyles through the loss of virulence and/or the acquisition of genes involved in tolerance to disinfectants (i.e. CC2 and CC6).

Clonogrouping is a PCR-based method that can be used to determine the main useful CCs, without the need for sequencing (Chenal-Francisque *et al.*, 2015).

7.13.1 Hypervirulent clonal complexes

The major hypervirulent clonal complexes, for example, CC1, CC4 and CC6, are strongly linked to clinical human and animal origins and appear to be associated with dairy products (Maury *et al.*, 2019). In general, the hypervirulent CCs (including CC1) appear to be better colonizers of the gut lumen and tissue than hypovirulent clones appear to be. CC1 in particular appears to be better adapted for within-host survival, persistence and faecal shedding. The better growth of the hypervirulent clones at 37 °C in the presence of salt, as compared with hypovirulent CCs, may account in part for the greater gut colonization capacity of hypervirulent CCs. This classification of hypovirulent and hypervirulent clones could help FBOs and risk managers take rapid action in the case of a food recall or a listeriosis outbreak.

Novel hypervirulent sublineages, isolated from listeriosis in goats, have recently been described. The isolates harbour both the *L. monocytogenes* LIPI-1 and a truncated LIPI-2 locus, encoding sphingomyelinase (SmcL), as well as other non-contiguous chromosomal segments from *L. ivanovii*. Interestingly, they also exhibit a unique wall teichoic acid structure essential for resistance to antimicrobial peptides, bacterial invasion and virulence (Yin *et al.*, 2019).

7.13.2 Hypovirulent clonal complexes

Hypovirulent clones usually contain a premature stop codon in one or more virulence factors, by far most commonly in *inlA*. These clones appear to be adapted to food-processing environments and have a greater prevalence of stress resistance and benzalkonium chloride tolerance genes and thus may show improved greater survival in the presence of sublethal concentrations of benzalkonium chloride. As a result, surface disinfection using benzalkonium chloride may actually provide hypovirulent CCs with a survival and/or growth advantage, favoring their greater persistence on surfaces or equipment treated with the chemical (Maury *et al.*, 2019). Examples of hypovirulent CCs include CC9 and CC121. The ST121/CC121 strains, which are among the most frequently found in food and the food-processing environment, contain a novel stress survival islet SSI-2, which supports survival under alkaline and oxidative stress conditions. In fact, ST121 strains have been reported to persist for months and even years in food-processing environments (Harter *et al.*, 2017).

7.14   EPIDEMIC CLONES

For *L. monocytogenes*, epidemic clones or ECs have been defined as i) a group of isolates that are genetically related and presumably of a common ancestor but are implicated in different geographically and temporally unrelated outbreaks, and ii) a clonal group that has been associated with more than one outbreak. Currently, ECs are defined mostly on the basis of multivirulence-locus sequence typing (Chen, Zhang and Knabel, 2007). The breakdown of ECs is as follows:

- ECI, ECII and ECIV – serotype 4b
- ECIII, ECV, ECVII – serotype 1/2a
- ECVI – serotype 1/2b

However, it should be noted that Cantinelli *et al.* (2013) found that using ECs to classify outbreak strains of *L. monocytogenes* is largely redundant and not as comprehensive as using MLST-defined clones. Therefore, we feel that providing EC information to risk managers would not be very informative.

7.15 SUBTYPE SPECIFIC RISK ASSESSMENTS AND VIRULENCE RANKING OF *L. MONOCYTOGENES*

One can use all of the above information to devise a comparative health risk evaluation of different strains of *L. monocytogenes* that can help to inform risk managers. A proposed ranking of the virulence of invasive strains of *L. monocytogenes* according to their lineage, possession of pathogenicity islands and whether or not they possess a full length or truncated internalin can be seen in Table 10. In addition, the potential advantages and disadvantages of using such a subtype specific risk assessment scheme are discussed in Table 11.

TABLE 10. Proposed virulence ranking of *L. monocytogenes* relevant to invasive listeriosis

Group	Description	Notes/confidence in classification rank	Examples of specific STs (and CCs)	Identification method
1	Lineage I strains with LIPI-1, 3 or 4 with a full length InlA	High confidence in classification, as most strains are virulent	ST1, ST3, ST4, ST5, ST6, ST73, ST87, ST54, ST619 (CC1, CC3, CC4, CC5, CC6, CC54, CC87)	(i) LIPI-1, 3, and 4 multiplex PCR; (ii) *hly/lisS/pts* multiplex PCR iii) WGS
2	Strain of any lineage with (i) complete and functional LIPI-1 and (ii) full length InlA (excluding strains in group 1)[a,b]	Could subgroup into 2a, 2b, and/or 2c (pending data on dose-responses for these groups)	ST1, ST3, ST4, ST5, ST6, ST73, ST7, ST8, ST9, ST37, ST87, ST101, ST121, ST155, ST204, ST1002 (CC1, CC3 -7, CC9[d], CC37, CC54, CC87, CC155, CC204)	*hly* and *inlA* SNP assay (or WGS)
3	Any *L. monocytogenes* strain with a truncated InlA[c]		ST8, ST9, ST121 (CC8, CC9[d], CC121)	*hly* and *inlA* SNP assay (or WGS)

[a] Risk managers may decide to include *L. innocua* with an intact LIPI-1 and an intact inlAB in group 2.
[b] Group 2 could potentially be subdivided into (i) Lineage I with a full length InlA; (ii) Lineage II with a full length InlA; and (iii) Lineages III and IV with a full length inlA; however, we are not aware of data that could be used to generate specific dose-response curves for each of these groups.
[c] Position and nature of the premature stop codon that leads to InlA truncation may affect virulence and strains with certain premature stops codons (such as those likely to revert to genes encoding full length InlA or those located in the 5' end) which could potentially be included in group 2; these strains are not necessarily avirulent.
[d] Some CCs (e.g. CC9) include strains both with and without a premature stop codon that leads to InlA truncation, and this may be listed in group 3 as well as in group 2.

TABLE 11. Potential advantages and disadvantages of developing a subtype specific risk assessment scheme

	Public health	Sustainability/resources	Cost/trade
Advantage	• Focuses public health and regulatory resources on highest human health risks, e.g. not wasting limited industry/public health resources • Encourages food industry and public health agencies to do WGS • Promotes further thinking into risk-based approaches for the control of *L. monocytogenes* in RTE foods	• Reduced number of recalls and hence decreased food waste and enhanced sustainability • Maintaining a secure and sufficient food supply	• Reduced recall costs • Fewer recalls that unnecessarily reduce consumer confidence (e.g. for healthy foods such as produce)
Disadvantage	• Public health impact may be difficult to assess • Possible unintended consequences, e.g. less *L. monocytogenes* testing due to higher costs • Can create difficult risk communication issues re the public, e.g. now allowing a *L. monocytogenes* containing food to be sold • A few cases of listeriosis may occur with food containing an extremely low virulence strain of *L. monocytogenes* • May create 2 different systems for doing risk assessments; one for low income and LMICs and another for upper-middle and high-income countries	• Might increase time to release product; increased time to resolve contamination issues	• Subtype specific risk management is cost intensive and may take away resources from higher needs, e.g. epi/surveillance • Could create trade barriers • Technically challenging, especially if multiple subtypes are found in the product

7.16 SUMMARY AND CONCLUSIONS

It is well recognized that almost all cases of listeriosis result from the consumption of a food containing high levels of *L. monocytogenes*. As such, control measures that prevent the occurrences of high levels of contamination at consumption have the greatest impact on reducing the rates of listeriosis. In fact, the vast majority of cases of listeriosis cases are associated with the consumption of foods that do not meet current Codex standards for *L. monocytogenes* in foods, whether the standard is "zero tolerance" (that is, not detected) or 100 CFU/g of *L. monocytogenes* (five samples for both). Most regulatory authorities globally would take action when finding any *L. monocytogenes* in an RTE food that supports growth, regardless of its strain characteristics.

In terms of genotypic/phenotypic markers of virulence, serotyping has limited value as a marker, and many developed countries are not doing classical serotyping anymore. Currently, the potential risk of a particular strain of *L. monocytogenes* causing severe illness in humans resulting from the consumption of food containing *L. monocytogenes* can be predicted using a combination of factors which could include i) molecular serotyping, ii) pathogenicity islands and their virulence genes (including presence of premature stop codons in virulence genes), iii) lineage, iv) sequence type, v) clonal complex type, vi) complex type based on WGS, and vii) host susceptibility. All of these factors can easily be deduced in-silico from the genomic sequence of the strain.

Currently, determining a subtype, although important, does not influence regulatory decision-making. However, we are proposing that a virulence ranking of *L. monocytogenes* obtained by determining and analysing subtyping data could be informative to improve risk assessments and thus make for better and more informed risk management decisions. However, after considering all the advantages and disadvantages of using a subtype specific risk assessment scheme, it is recommended that the control of *L. monocytogenes* globally should continue to use an approach that does not consider subgroups (ST/CC) of *L. monocytogenes* but allows risk managers in some countries to use *L. monocytogenes* subtype information to inform risk management. As we learn more about the key virulence factors/markers in *L. monocytogenes*, how they function and how they are controlled, and as WGS becomes more widely used by member countries around the world, regulatory authorities may be able to use this information to make better risk management decisions, particularly since such additional information can be obtained by accessing public *L. monocytogenes* genetic sequence databases that exist worldwide.

Some hypothetical example scenarios where using this approach could be beneficial are presented below. The examples below are more pertinent for those member countries who do not have a "zero tolerance" for all foods.

Example 1

SITUATION: *L. monocytogenes* is found in an RTE food that is characterized as not allowing greater than 100 CFU/g of *L. monocytogenes* at the end of shelf-life. The levels of *L. monocytogenes* are tested and found to be about 90 CFU/g. The strain is identified as a serotype 4b, belonging to lineage I, CC1, and it contains the LIPI-1, LIPI-3 and LIPI-4 pathogenicity islands, i.e. a potentially hypervirulent strain.

DECISION: RECALL/WITHDRAW OR NOT? POTENTIAL DECISION: **RECALL/WITHDRAW**

RATIONALE: The hypothetical strain that has been found is thought to be hypervirulent. Even though *L. monocytogenes* will not grow to greater than 100 CFU/g in this particular RTE food, and the tested levels confirm this, competent authorities may decide that the risks are too great in this situation and recall the food from the marketplace.

Example 2

SITUATION: *L. monocytogenes* is found in a low moisture food at levels slightly greater than 100 CFU/g. It is found to be a serotype 1/2c, lineage II strain, which has a truncated internalin and is therefore hypovirulent.

DECISION: RECALL/WITHDRAW OR NOT? POTENTIAL DECISION: **NO RECALL/WITHDRAWAL**

Further question: Does the decision change if targeted to at-risk populations?

RATIONALE: The hypothetical strain that has been found in the LMF is known to be hypovirulent. Even though the levels are just above 100 CFU/g, competent authorities may decide that the risk is worth taking because it is an LMF in low supply and it is a staple food in the country.

If the food was specifically targeted to at-risk populations, e.g. hospitalized individuals, the risk level would increase to the extent that the competent authorities decide that the risk is too great and they recall the food from the marketplace.

Example 3

SITUATION: An atypical strain of *L. innocua* is found in an RTE food supporting growth of *Listeria* at levels less than 100 CFU/g. It is found to be hemolytic and contains a full LIPI-1 pathogenicity island.

DECISION: RECALL/WITHDRAW OR NOT? POTENTIAL DECISION: **RECALL/WITHDRAW**

RATIONALE: Since this is a food that can support the growth of *L. innocua* and it appears to contain some virulence factors characteristic of *L. monocytogenes*, the risk competent authorities may decide that even though *L. innocua* is normally considered non-pathogenic, in this case it has some characteristics of the pathogen *L. monocytogenes* and the risk is too great, and thus they would recall the food from the marketplace.

Example 4

SITUATION: A strain of *L. ivanovii* is found in an RTE food supporting its growth. It is found to be present at levels of 10^6 CFU/g.

DECISION: RECALL/WITHDRAW OR NOT? POTENTIAL DECISION: **RECALL/WITHDRAW**

RATIONALE: This is a food that can support the growth of *L. ivanovii*, and this species, although not a common cause of human disease, has caused the occasional human illness. Levels of 106 CFU/g also signify that something went wrong, and either there were supplier issues or inadequate GMPs in the plant. Thus, in this case, the competent authorities decide that the risk is too great for the public and do a recall.

7.17 RECOMMENDATIONS

- Key new information has emerged on *L. monocytogenes* virulence differences that should be considered in future *L. monocytogenes* risk assessments, as well as in *L. monocytogenes* risk management.
- However, considering all the advantages and disadvantages, *L. monocytogenes* control globally should continue to use an approach that does not consider *L. monocytogenes* subgroups, while allowing risk managers in some countries to use *L. monocytogenes* subtype information to inform risk management.
- Although the food matrix and its previous supply-chain environment could influence the expression of *L. monocytogenes* virulence genes, we feel that there is not enough data to include this as a variable in dose-response models.
- Parameters for the dose-response relationship for *L. monocytogenes* need to be updated.
- A separate advisory group should be tasked with deciding whether the current exponential dose-response model for *L. monocytogenes* is still adequate.

8

Exposure assessment

8.1   NEW TRENDS AND INFORMATION SINCE THE LAST MICROBIOLOGICAL RISK ASSESSMENT IN 2004

The FAO/WHO Risk assessment of *L. monocytogenes* in RTE foods (MRA 5) (FAO and WHO, 2004a, 2004b) did not consider the full production to consumption continuum but focused on the retail to consumption stage. The assessment focused on four RTE products, which differed in their ability to sustain the growth and/or the presence of a CCP for the control of *L. monocytogenes*; cold-smoked salmon, pasteurized milk, ice cream and fermented meats are examples.

The following section pertains to factors warranting updates to the exposure assessment with a focus on new information about the prevalence and growth potential in different types of food, as well as on biological strain diversity within *L. monocytogenes* and the possible contribution of these factors to the risk of exposure.

Since the publication of the FAO/WHO MRA 5 in 2004, outbreaks of listeriosis have occurred in a range of products not previously thought to be at-risk products. Examples include cantaloupe or rock melon, caramel apples, frozen peas and corn, and produce and ice cream used in milk shakes. The ability of the bacterium to grow in RTE foods during storage is one of the most important determinants of risk as shown in Table 12 (see also Chapter 7 Hazard characterization).

TABLE 12. Assessment of potential *L. monocytogenes* exposure as influenced by consumption habits, growth potential of the bacterium, risk per serving and prevalence of the bacterium in different types of RTE foods with a special focus on produce.

Product	Consumption	Growth and/or survival potentiala	Risk per serving	Prevalence	Example of foods/characteristics
Milk	+++	+++[b]	+	-	Pasteurization
Ice cream	+++	0/+++ if stored at > 0°C	-	-	Frozen storage (unintended refrigerated storage)
Fermented meat	+	-	-	+++	Summer sausage
RTE seafood	+	+++	+++	+++	Cold-smoked salmon, gravid fish, shrimp, formulation not inhibitory
Cold-cut meats	+++	+++	+++	+++	Ham, pastrami, cured beef, formulation not inhibitory
Soft cheeses surface ripened with mould	++	+++	-	++	Brie, camembert
Unripened soft cheeses	+	+++	+	++	Hispanic style, queso fresco
Soft cheeses (blue veined)	+	+/+	-	+	Blue cheeses
Leafy greens	+++	+(whole)/++ (cut)	-	-/+	Arugula, iceberg, lettuce, endive, spinach, kale
Cut vegetables	++	0/+	-	-	Carrots, celeriac, onions
Uncooked vegetables	++	++	-	+	Peppers, bean sprouts, avocado, broccoli, white cabbage, cauliflower, cucumber, celery
Fungi	++	++	-	+	Mushrooms, enoki
Composite RTE salads	+++	++	-	+	No added mayonnaise/dressing
Frozen vegetables	+	0/++ if stored at > 0°C	-	+	Corn (cooked), peas (blanched)

(cont.)

Fruits (pH ≥ 4.5)	++	++	-	+	Melons (cantaloupe/rock, honeydew, water)
Fruits (pH < 4.5)	++	0/+	-	+	Strawberries, raspberries, pears, apples, grapes

[a] Growth within the shelf-life (information based on Beuchat *et al.* 2002; Hoelzer *et al.*, 2012b; Lokerse *et al.*, 2016; Marik *et al.*, 2020; Mejlholm *et al.*, 2010; Oladimeji *et al.*, 2019; Østergaard *et al.*, 2014; Zeng *et al.*, 2014; Ziegler *et al.*, 2019).
[b] Legend for growth potential: 0 – no growth, + up to a 0.5 log increase, ++ up to a 2-log increase, +++ more than a 2-log increase.
[c] The risk per serving is indicated as either high risk (+++), moderate risk (+), or low risk (-) (FDA/FSIS, 2003).

In many types of RTE produce, the growth potential of *L. monocytogenes* is minimal with the bacterium being unlikely to reach high levels that pose a significant risk before the product has spoiled. This inhibition or competition from the endogenous spoilage microbiota that decreases the growth rate and/or maximum population is known as the Jameson effect (Ross *et al.*, 2000). However, the role of the indigenous or background microflora in reducing the growth potential of *L. monocytogenes* remains to be fully elucidated and can vary with food type.

8.2 ENVIRONMENTAL FITNESS

Since the FAO/WHO risk assessment on *L. monocytogenes* in RTE foods in 2004, substantial progress has been made in quantifying the growth limits or boundaries for *L. monocytogenes* in regard to various intrinsic parameters including pH, a_w, organic acids, salt, nitrate, phenolics, and smoke compounds. Some of the cardinal parameters, which can be used to characterize *L. monocytogenes* growth, are shown in Table 4, Section 5.3.2.

The field of predictive modelling has led to the development of publicly available software such as the Food Spoilage and Safety Predictor (FSSP) (http://fssp.food. dtu.dk/Help/Index.htm), which considers up to 14 different parameters related to the characteristics of the food (pH, salt, organic acids, nitrite, temperature, and so on) including a model for the interactions between lactic acid bacteria and *L. monocytogenes*. These models have been validated for RTE meat and seafood products (Mejlholm *et al.*, 2010) as well as cottage cheese (Østergaard *et al.*, 2014). An expanded version of FSSP is expected in 2021 that will include additional cheese products. However, currently no models are available to accurately predict the growth of *L. monocytogenes* in RTE produce or composite RTE salads made from combinations of vegetables, leafy greens, meat and seafood. Another useful tool for predictive modelling is the ComBase Predictive Models, which are a collection of software tools based on a very large amount of data, which can be used to predict the growth or inactivation of microorganisms as a function of environmental factors such as temperature, pH and water activity in broth. These models can help to inform the design of food safety risk management plans and assess potential microbiological risks in food (https://www.combase.cc/index.php/en/).

8.3 STRAIN DIVERSITY

Listeria monocytogenes strains, which have significantly different growth potential or tolerance to stress conditions including biocides, may select for their survival and dominance in environmental niches (Abee *et al.*, 2016). These strains are now receiving increasing attention, with their environmental fitness being one of the explanations for the long-term persistence of certain clones or types in factory environments (Ferreira *et al.*, 2014; Stoller *et al.*, 2019).

In the context of exposure assessment, better survival and/or growth potential may translate into greater numbers of *L. monocytogenes* being present in RTE foods at the time of consumption. Table 13 below lists some of the environmental factors for which strain dependent variations have been examined.

TABLE 13. Examples of studies of naturally occurring strain diversity in tolerance to environmental stress factors

	Factor	Diversity (increased growth/survival/tolerance)	Reference
Extrinsic	Low temperature <5 °C	Absence of PMSCs[a] in *sig*B operon, full length *inl*A, lower motility	Cordero *et al.*, 2016; Hingston *et al.*, 2017
Intrinsic	Salt (osmotic)	Full length *inl*A, CC2 and CC11 more tolerant to 6% salt, lineage I more tolerant to 9% salt than lineage 2	Hingston *et al.*, 2017; Aalto-Araneda *et al.*, 2020
	pH	SSI-1, mutations in *rpsU*, plasmid harbourage,	Ryan *et al.*, 2010; Metselaar *et al.*, 2015; Hingston *et al.*, 2017
	a_w	Absence of PMSCs in *sig*B operon	Hingston *et al.*, 2017
	Nisin	Mutations in *rsb*U, *pbpb3* (lmo0441)	Wambui *et al.*, 2020

(cont.)

	Benzalkonium chloride (QAC)	*erm*C (pLMST6), *qac*H-Tn*6188*, *brc*ABC, *qac*C	Hurley *et al.*, 2019; Kropac *et al.*, 2019; Stoller *et al.*, 2019
Environmental	Heat	Mutations in *rps*U, *cts*R	Den Besten *et al.*, 2018
	High pressure processing	Mutations in *cts*R	Van Boeijen *et al.*, 2010

ᵃ Premature stop codons.

Increased stress resistance generally comes with a trade-off where the variant will exhibit a slower growth rate under non-stress conditions. Since predictive microbiology models assume that all *L. monocytogenes* strains have the same "average" tolerance (± some variations), more research is clearly needed to determine if the models should include terms that account for stress variant strains, including integration of strain genetic information in the exposure and risk assessment (Fritsch *et al.*, 2018; Njage *et al.*, 2020).

8.4   CONCLUSION AND RECOMMENDATIONS

In the FAO/WHO MRA5 report (FAO and WHO, 2004b), a retail-to-consumer risk assessment was done for milk, frozen ice cream, fermented meat, and cold smoked salmon. Since fresh produce is only minimally processed and has been increasingly involved in foodborne outbreaks, we recommend that production to consumption risk assessments be considered for these types of products. Based on our discussions and new knowledge about *L. monocytogenes* prevalence, growth, new dose-response, other factors of importance for growth or survival and outbreaks of *L. monocytogenes*, the expert group recommends that the following produce types be chosen as the focus for the next risk assessment(s):

- leafy greens
- cantaloupe/rock melon
- frozen vegetables (e.g. peas, corn)

One could also consider updating the risk assessment done on RTE seafood that allows for the growth of *L. monocytogenes* in salmon or halibut, e.g. gravid (sugar-salt marinated).

Nonetheless, it is hypothesized that the main issue remains *L. monocytogenes* colonization of the processing environment due to poor sanitation and lack of hygienic design. There may also be a climate change dimension that needs to be included in future *Listeria* risk assessments that incorporate the production to consumption continuum.

New information is appearing regarding strain variants, where the genetic make-up confers enhanced tolerance to disinfectants, low water activity, heat, or increased growth rates at low temperatures compared to other strains. Whether these variants have an advantage in the food ecosystem or niche depends on correlations among the different food stress factors encountered. These variations in growth or survival appear to be complex and are presently poorly understood, making their inclusion in risk assessments and risk management decisions difficult. Continued research should help to determine if inclusion of these strain variants is warranted in future *Listeria* exposure models.

Lastly, it should be noted that discussions of factors warranting updates to the exposure assessment focused on new information about the prevalence, growth potential in different types of food as well as on biological strain variability within *L. monocytogenes* and on the possible contribution of these factors to the risk of exposure. It is a clear limitation that the impact of dietary or consumer habits on the potential exposure to *L. monocytogenes* was not covered. Future exposure assessments should examine this limitation as well as some of the considerations raised in Chapter 9 of this report.

9

Other considerations

To reduce foodborne listeriosis, risk communication strategies must be developed to clearly communicate risk factors associated with product storage, shelf-life and appropriate consumption of RTE foods by at-risk consumers. Effective messaging that targets pregnant women has been achieved, but how can effective public health messaging be targeted to other groups at risk? Quantitative modelling by EFSA (2018) suggested that more than than 90 percent of cases of invasive listeriosis were caused by the ingestion of RTE food containing > 2 000 CFU of *L. monocytogenes*/g, and that one-third of the cases are due to growth of *L. monocytogenes* in the consumer phase.

EFSA recommends a focus on strategies that increase awareness of listeriosis and encourage appropriate food handling among susceptible at-risk populations. Recent outbreaks in the United States of America and Europe have shown an increase of listeriosis in older adults (> 60 years of age). Behavioural risk factors identified in older adults included a lack of adherence to use-by dates and ineffective refrigerated storage of RTE foods. Targeted messaging specifically for older consumers is recommended (Evans and Redmond, 2014). Additionally, data specific to older adult risk factors associated with listeriosis are lacking and require further identification. For instance, Kvistholm Jensen *et al.* (2017) examined the use of proton pump inhibitors (PPIs) among individuals in Denmark. Following adjustment for comorbidities and additional confounding variables, a 2.8-fold increased risk for listeriosis risk was found to be associated with the use of PPIs.

Gillespie *et al.* (2010), in an examination of cases of listeriosis in England occurring between 2001 and 2007, found an association with neighbourhood deprivation. For all patient groups, listeriosis incidence was highest in the most deprived areas of England when compared with the most affluent, and those with cases of listeriosis were more likely to purchase foods from convenience stores or from local services (bakers, butchers, fishmongers and greengrocers) in comparison with the general population. When patient age was available for non-pregnancy associated cases, 76 percent of the cases were aged 60 years or older. With increased life expectancy and rising food prices, food insecurity could become an increasingly important driver for foodborne disease generally, and listeriosis specifically, in the future.

Newsome *et al.* (2014) examined applications and perceptions of date labelling of food from a global perspective. The many variations in date labelling (use by, consume by, best before, expires on) contribute to confusion and misunderstanding regarding how the dates on labels relate to food quality or safety. This confusion and misunderstanding were found to be significant contributors to food waste. In the United States of America, it is estimated that 90 percent of Americans prematurely discard food, and as much as 40 percent of the United States of America's unused food supply is discarded annually due to confusion regarding food date labels. Efforts to provide education regarding the meaning of date labelling terms, the importance of shelf-life limitation for some products, and temperature control, availability and understanding of food storage guidance, as well as safe handling methods could significantly impact not only a reduction in food waste, but also improved food safety.

Individuals who may be susceptible to listeriosis due to age, pregnancy or immunosuppressive conditions should be advised to avoid high-risk foods such as raw milk soft cheeses and instead seek lower risk alternatives. In the case of frozen, non-RTE vegetables, adherence to package instructions for cooking at recommended temperatures prior to consumption is important, and consistency in label instructions may improve adherence to recommended guidance. Consumer interest in high quality nutritious foods to support a healthy lifestyle has resulted in increased consumption of produce and concomitant exposure to *Listeria*. Fresh produce has emerged as a major source of foodborne illness outbreaks linked to *L. monocytogenes*, and thus public health risk communication messaging should make the public aware of this, while still stressing the importance of produce as part of a healthy diet.

Additional factors that might influence the risk of listeriosis could also be related to i) social, administrative, and economical issues; ii) access to healthcare and surveillance of infectious diseases; iii) behavioral and cultural factors; iv) burden of other diseases or underlying health issues; and v) other factors. It should also be noted that in some countries, foodborne illnesses such as listeriosis may not be a priority, but this does not mean that they should be neglected. Examples related to factors that could directly or indirectly influence the risk of listeriosis are provided in Table 14.

TABLE 14. Other considerations for future work and research – additional factors that might influence the risk of listeriosis in low- and middle-income countries (LMICs)

Additional factors that might influence the risk of listeriosis in LMICs	Examples
Social, administrative and economic issues	• Unstable political conditions leading to lack of basic services and/or food • Economic issues leading to lack of basic services and/or food • Censorship of health data: gaps in the information of the true burden of listeriosis, causing this illness to be neglected resulting in non-existent or insufficient work on risk communication, surveillance, notification of cases, etc. • Immigration, potentially leading to a lack of access to food and basic hygiene due to the precarious situation of this population and their habitats • Difficulties in the diagnosis of listeriosis due to lack or cost of laboratory supplies and/or equipment for detecting *L. monocytogenes*
Access to healthcare and surveillance of infectious diseases	• Access to healthcare • Weak or non-functional healthcare systems • Lack of medical supplies • Non-existent or incomplete surveillance programme (where the actual burden of listeriosis is unknown)
Behavioral and cultural factors	• Consumption of high-risk foods as part of the local culture (unpasteurized milk or dairy products, for example) • Traditions or rituals involving consumption of potentially high-risk foods (such as raw meats) • Religious beliefs
Burden of other diseases or underlying health issues	• HIV • Tuberculosis • Malnourishment • Diabetes • Treatments for chronic diseases • Untreated underlying diseases
Other factors	• Lack of information about the burden of listeriosis • Access to sanitation • Access to clean water for drinking and cooking

10

Overall conclusions

Overall conclusions for each of the Chapters are summarized below. In addition, examples from literature for seven important foodborne listeriosis outbreaks linked to dairy, produce and meat products were also developed by the expert group, and details on the key learnings from these outbreaks are provided in Annexes A1.1 to A1.8.

10.1 SOURCE ATTRIBUTION AND BURDEN OF DISEASE

The global burden of foodborne disease associated with listeriosis

- The WHO FERG estimated that data on the burden of listeriosis in 2010 represented only 48 percent of the world's population. Some individual countries have estimated their own burden of listeriosis, but these studies were limited to high-income regions.
- The incorporation of new data on the incidence of invasive and, if available, feasible and appropriate, non-invasive forms of listeriosis in LMICs (Africa, Latin America, South Asia), through a systematic review of peer-reviewed studies and national surveillance, would make these estimates more globally representative and more precise.
- The relevance of gastroenteritis on the listeriosis burden of disease is not fully understood but is currently considered to be minimal compared to invasive listeriosis.

- At this point we do not recommend surveillance for gastrointestinal listeriosis. Instead, focus should be restricted to the development and implementation of effective surveillance systems for invasive listeriosis.

Source attribution of foodborne listeriosis

- Several national listeriosis source attribution studies have been reported using various methods (e.g. expert opinion, genomic data analysis, risk assessment). These studies are, however, limited to high-income regions, and no global listeriosis source attribution study exists to date.
- The expert group underlined the necessity for having a harmonized food classification scheme (ontology) at the international level for the categorization and typing of foods considered in source attribution studies.
- Source attribution studies based on systematic reviews of outbreak data and case-control data of sporadic cases should be set up at the global level.
- The public health impact and the number of listeriosis cases occurring in countries are much larger than the major outbreaks alone. Outbreaks can give useful information and learning opportunities, but many sporadic cases are still occurring due to the consumption of contaminated RTE foods. More outbreaks, some due to novel foods, can and will be detected, but these outbreaks will still comprise the minority of all cases.
- However, to the extent that sporadic cases are transmitted by the same types of foods associated with outbreaks, outbreak investigations that lead to enhanced regulatory and food industry controls can be expected to reduce or control the incidence of listeriosis.

Susceptible populations

- We recommend that public risk communication focus on informing identified risk groups about their relative susceptibility as well as about foods that carry a particularly high risk of containing *L. monocytogenes*.
- Based on susceptibility, consideration should be given to a dose-response model for the three main subpopulations:
  » less susceptible subpopulation (i.e. the general population under 65 years old with no known risk factors for listeriosis or comorbidities);
  » susceptible subpopulation: pregnant women and their newborns as well as adults aged 65 or older; and
  » very susceptible subpopulation (for example, people with weakened immune systems such as HIV-infected individuals, cancer patients and organ transplant patients).

- There is a need to update the relative values for risk factors and comorbidities used in dose-response models, based on recent large cohort studies, newly recognized risk factors (high fat diet, etc.) and data collected from global burden estimates.

10.2 MONITORING

- Internationally, microbiological criteria for *L. monocytogenes* are either based on a "zero tolerance" (not detected in the sample tested) approach in RTE foods that support the growth of *L. monocytogenes* or on an approach that permits low levels in those foods that do not support the growth of the organism.
- Monitoring of finished products performs a verification role but provides only limited assurance of the safety of an RTE food.
- Scientific evidence demonstrates that the majority of the effort should be focused on the proactive management and control of *L. monocytogenes* by implementing a comprehensive food safety management system (FSMS). A key component of such a programme is rigorous environmental monitoring to assess incursions and the ongoing presence of *Listeria* spp. in food-processing facilities, especially in those zones in close proximity to food-contact surfaces.
- Food businesses should adopt a risk-based approach which incorporates effective process control, environmental monitoring, and appropriate final product testing to manage potential risks.
- Regulatory agencies are encouraged to use a combination of finished product testing and environmental monitoring to verify proper implementation of *L. monocytogenes* control strategies by food businesses.
- Best practices involve subtyping (preferably WGS) of *L. monocytogenes* isolates obtained through monitoring and surveillance to support the creation of databases of human, food and environmental isolates. WGS of isolates assists with tracking and tracing outbreaks, can assess the potential for virulence, and determine whether strains have established residence in food-processing facilities.

Monitoring and surveillance: how it is done, by whom and at what frequency
- The approach varies between countries, with the CAs in some jurisdictions actively monitoring *L. monocytogenes* in RTE foods, while others evaluate company monitoring through audit processes.
- Where RTE food is found to exceed microbiological limits, a recall is typically required. However, in some countries, the confirmation of *L. monocytogenes* on food contact surfaces can also be a reason to initiate a recall.

- Regulatory strategies should encourage aggressive environmental monitoring to eliminate sources of *L. monocytogenes*. Microbiological criteria (for example, zero tolerance) can negatively impact the implementation of FSMS and particularly the way environmental monitoring programme are employed and reported.

Other issues: reflections on issues raised in examples from literature – politics, health care and resources, labelling (especially for frozen food that is to be cooked and may be consumed raw)

- Risk reduction is the responsibility of the food industry, governments and consumers.
- Modelling scenarios for the production, handling and consumption of RTE foods such as melons and deli meat could reveal potential areas where public health improvements could be made.
- The emergence of fresh produce as an important source of foodborne listeriosis has now become a global public health problem which may require consideration of risk communication to the public. This could include, for example, that high-risk groups i) need to pay attention to date labelling on fresh-cut packaged leafy greens; ii) should not eat raw or lightly cooked sprouts of any kind; and iii) should eat cut melon right away or refrigerate it and should discard cut melons left at room temperature for more than four hours.
- Education to consumers about the safety of produce, especially for those who are most at risk, should include other produce such as frozen vegetables that are sometimes consumed without further heating.
- Additional factors that may influence the risk of listeriosis in certain regions of the world, such as LMICs, include (i) social, administrative, and economic issues; (ii) access to healthcare and surveillance of infectious diseases; (iii) behavioural and cultural factors; and (iv) burden of other diseases or underlying health issues.
- In order to reduce foodborne listeriosis, risk communication strategies must be developed to clearly communicate risk factors associated with food storage, shelf-life, and appropriate consumption of RTE foods by vulnerable consumers.
- The provision of targeted education regarding the meaning of date labelling terms, the importance of shelf-life limitation for some products, temperature control, availability and understanding of food storage guidance as well as safe handling methods will improve food safety and could significantly reduce food waste.

Other issues: comorbidities and their impact

- Increasing age amplifies the likelihood of death in listeriosis cases. This reflects the fact that comorbidities, risk factors and immunodeficiencies increase with age. Unfortunately, information on comorbidities is often lacking from epidemiological studies of listeriosis, and more research into the underlying causes of increased susceptibility to listeriosis is needed.

- Risk-benefit considerations should be used to identify the trade-off between reduced availability of nutritious foods due to unnecessary recalls and any potential risk.

10.3 LABORATORY METHODS

- The use of harmonized and validated laboratory methods is critical to obtain reliable results and strengthen surveillance systems.

- Culture and isolation should be the foundation of laboratory methods, so that isolates are available for further characterization. Chromogenic media are particularly useful for isolating *L. monocytogenes* either directly or following enrichment. Consideration should also be given to the recovery of sublethally injured cells from both food and environmental samples.

- Metagenomic analysis of enrichments could be a promising approach for real-time detection. Several public health relevant markers such as antimicrobial resistance, biocide tolerance, as well as whole genome sequencing of *L. monocytogenes* isolates can provide additional valuable sources of information for monitoring, risk assessments, and recalls and outbreak investigations, leading to better informed management decisions.

10.4 HAZARD CHARACTERIZATION

- Important new information has emerged on *L. monocytogenes* virulence differences that should be considered in future risk assessments, as well as in risk management.

- However, considering all the advantages and disadvantages, the global control of *L. monocytogenes* should continue to use an approach that does not consider *L. monocytogenes* subgroups, while allowing risk managers in some countries to use *L. monocytogenes* subtype information to inform risk management decisions.

- Although the food matrix and its previous supply-chain environment could influence the expression of *L. monocytogenes* virulence genes, the expert group felt that there is not enough data to include this as a variable in dose-response models.

- Parameters for the dose-response relationship for *L. monocytogenes* need to be improved to give better information for risk managers.
- A separate advisory group should be tasked with deciding whether the current exponential dose-response model for *L. monocytogenes* is still adequate.

10.5 EXPOSURE ASSESSMENT

- The expert group recommends that risk assessments should be done by FAO/WHO (JEMRA), as this ensures more stakeholders are being heard. Having international experts collaborating will give the risk assessment more global relevance, credibility, transparency, and better visibility world-wide. In other words, it lifts the risk assessment up to an international level.
- Based on discussions and new knowledge about *L. monocytogenes* prevalence, growth, dose response, outbreaks, as well as other factors of importance for the growth/survival of *L.* monocytogenes, the expert group recommends that the following produce types be chosen as the focus for the next risk assessment(s):
  - » leafy greens
  - » cantaloupe/rock melon
  - » frozen vegetables (for example peas, corn)

  One could also consider updating the risk assessment done on:
  - » RTE seafood that allows for the growth of *L. monocytogenes,* for example, gravad (sugar/salt marinated) salmon/halibut.
- New information is appearing regarding *L. monocytogenes* strain variants; where compared to other strains, the genetic make-up of the variants appears to confer enhanced tolerance to disinfectants, low water activity, heat, or increased growth rates at low temperatures. Whether these variants have an advantage in the food ecosystem or niche depends on the correlation among the different food stress factors encountered. At this time, the reported variations in growth or survival of *L. monocytogenes* appear complex or there is insufficient knowledge, making inclusion in risk assessments and risk management decisions difficult. However, research on strain variants should continue to expand our knowledge and be used for possible future inclusions in the modelling of *L. monocytogenes* exposure.
- In the previous MRA5 (FAO and WHO, 2004b), it was decided to do a retail-to-consumer risk assessment only for milk, frozen ice cream, fermented meat and cold-smoked salmon. Since fresh produce is minimally processed, we recommend that farm-to-fork risk assessments should be considered for

these types of products. Furthermore, we hypothesize that the main issue remains *L. monocytogenes* colonization of the processing environment due to poor sanitation and lack of hygienic design. There may also be a climate change dimension that needs to be included in future risk assessments of *L. monocytogenes* that incorporates the entire farm-to-fork approach. We also recommend farm-to-fork risk assessments that include climate change, including shorter term weather factors, as well as precision agricultural practices.

References

Aalto-Araneda, M., Pöntinen, A., Pesonen, M., Corander, J., Markkula, A., Tasara, T., Stephan, R. & Korkeala, H. 2020. Strain variability of *Listeria monocytogenes* under NaCl stress elucidated by a high-throughput microbial growth data assembly and analysis protocol. *Applied and Environmental Microbiology*, 86(6): e02378–19. https://doi.org/10.1128/AEM.02378-19

Abee, T., Koomen, J., Metselaar, K. I., Zwietering, M.H. & den Besten, H.M. 2016. Impact of pathogen population heterogeneity and stress-resistant variants on food safety. *Annual Review of Food Science and Technology*, 7: 439–456. https://doi.org/10.1146/annurev-food-041715-033128

Alali, W.Q. & Schaffner, D.W. 2013. Relationship between *Listeria monocytogenes* and *Listeria* spp. in seafood-processing plants. *Journal of Food Protection*, 76(7): 1279–1282. https://doi.org/10.4315/0362-028X.JFP-13-030

Althaus, D., Jermini, M., Giannini, P., Martinetti, G., Reinholz, D., Nüesch-Inderbinen, M., Lehner, A. & Stephan, R. 2017. Local outbreak of *Listeria monocytogenes* serotype 4b sequence type 6 due to contaminated meat Pâté. *Foodborne Pathogens and Disease*, 14(4): 219–222. https://doi.org/10.1089/fpd.2016.2232

Amato, E., Filipello, V., Gori, M., Lomonaco, S., Losio, M.N., Parisi, A., Huedo, P., Knabel, S.J. & Pontello, M. 2017. Identification of a major *Listeria monocytogenes* outbreak clone linked to soft cheese in Northern Italy - 2009-2011. *BMC Infectious Diseases*, 17(1): 342. https://doi.org/10.1186/s12879-017-2441-6

Angelo, K.M., Conrad, A.R., Saupe, A., Dragoo, H., West, N., Sorenson, A., Barnes, A., Doyle, M., Beal, J., Jackson, K. A., Stroika, S., Tarr, C., Kucerova, Z., Lance, S., Gould, L.H., Wise, M. & Jackson, B.R. 2017. Multistate outbreak of *Listeria monocytogenes* infections linked to whole apples used in commercially produced, prepackaged caramel apples: United States, 2014-2015. *Epidemiology and Infection*, 145(5): 848–856. https://doi.org/10.1017/S0950268816003083

Augustin, J.C. & Carlier, V. 2000. Mathematical modelling of the growth rate and lag time for *Listeria monocytogenes*. *International Journal of Food Microbiology*, 56: 29–51.

Bakke, M., Suzuki, S., Kirihara, E. & Mikami, S. 2019. Evaluation of the total adenylate (ATP + ADP + AMP) test for cleaning verification in healthcare settings. *Journal of Preventive Medicine and Hygiene*, 60(2): E140–E146. https://doi.org/10.15167/2421-4248/jpmh2019.60.2.1122

Batz, M.B., Hoffmann, S. & Morris, J.G., Jr. 2012. Ranking the disease burden of 14 pathogens in food sources in the United States using attribution data from outbreak investigations and expert elicitation. *Journal of Food Protection*, 75(7): 1278–1291. https://doi.org/10.4315/0362-028X.JFP-11-418

Batz, M., Hoffmann, S. & Morris, J.G., Jr. 2014. Disease-outcome trees, EQ-5D scores, and estimated annual losses of quality-adjusted life years (QALYs) for 14 foodborne pathogens in the United States. *Foodborne Pathogens and Disease*, 11(5): 395–402. https://doi.org/10.1089/fpd.2013.1658

Beaufort, A., Cornu, M., Bergis, H., Lardeux, A.L. & Lombard, B. 2014. *EURL Lm Technical guidance document for conducting shelf-life studies on Listeria monocytogenes in ready-to eat foods.* Version 3. Community Reference Laboratory for *Listeria monocytogenes*. www.fsai.ie/uploadedFiles/EURL%20Lm_ Technical%20Guidance%20Document%20Lm%20shelf-life%20studies_V3_2014-06-06%20(2).pdf

Belias, A., Brassill, N., Roof, S., Rock, C., Wiedmann, M. & Weller, D. 2021. Cross-validation indicates predictive models may provide an alternative to indicator organism monitoring for evaluating pathogen presence in Southwestern US agricultural water. *Frontiers in Water*, 3: 693631. https://doi.org/10.3389/frwa.2021.693631

Beuchat, L.R. 2002. Ecological factors influencing survival and growth of human pathogens on raw fruits and vegetables. *Microbes and Infection*, 4(4): 413–423. https://doi.org/10.1016/s1286-4579(02)01555-1

Brent, P. 2012. Risk factors for listeriosis in Australia. *Epidemiology and Infection*, 140(5): 878. https://doi.org/10.1017/S0950268811001117

Brown, E., Dessai, U., McGarry, S. & Gerner-Smidt, P. 2019. Use of whole-genome sequencing for food safety and public health in the United States. *Foodborne Pathogens and Disease*, 16(7): 441–450. www.liebertpub.com/doi/10.1089/fpd.2019.2662

Buchanan, R.L., Gorris, L.G.M., Hayman, M.M., Jackson, T.C. & Whiting, R.C., 2017. A review of *Listeria monocytogenes*: An update on outbreaks, virulence, dose-response, ecology, and risk assessments. *Food Control*, 75: 1–13. https://doi.org/10.1016/j.foodcont.2016.12.016

Burall, L.S., Grim, C.J. & Datta, A.R. 2017. A clade of *Listeria monocytogenes* serotype 4b variant strains linked to recent listeriosis outbreaks associated with produce from a defined geographic region in the US. *PloS One*, 12(5): e0176912. https://doi.org/10.1371/journal.pone.0176912

Cantinelli, T., Chenal-Francisque, V., Diancourt, L., Frezal, L., Leclercq, A., Wirth, T., Lecuit, M. & Brisse, S. 2013. "Epidemic clones" of *Listeria monocytogenes* are widespread and ancient clonal groups. *Journal of Clinical Microbiology*, 51(11): 3770–3779. https://doi.org/10.1128/JCM.01874-13

Cassini, A., Colzani, E., Pini, A., Mangen, M.-J.J., Plass, D., McDonald, S.A., Maringhini, G. *et al.* 2018. Impact of infectious diseases on population health using incidence-based disability-adjusted life years (DALYs): results from the Burden of Communicable Diseases in Europe study, European Union and European Economic Area countries, 2009 to 2013. *Eurosurveillance*, 23(16): 17–00454. https://doi.org/10.2807/1560-7917.ES.2018.23.16.17-00454

CDC (Centers for Disease Control and Prevention [United States of America]). 1998. Multistate outbreak of listeriosis -- United States of America, 1998. MMWR 47: 1085–6.

Chalenko, Y., Kalinin, E., Marchenkov, V., Sysolyatina, E., Surin, A., Sobyanin, K. & Ermolaeva, S. 2019. Phylogenetically defined isoforms of *Listeria monocytogenes* invasion factor InlB differently activate intracellular signaling pathways and interact with the receptor gC1q-R. *International Journal of Molecular Sciences*, 20(17): 4138. https://doi.org/10.3390/ijms20174138

Charlier, C., Perrodeau, É., Leclercq, A., Cazenave, B., Pilmis, B., Henry, B., Lopes, A. *et al.* 2017. Clinical features and prognostic factors of listeriosis: the MONALISA national prospective cohort study. *The Lancet Infectious Diseases,* 17(5): 510–519. https://doi.org/10.1016/S1473-3099(16)30521-7

Charlier, C., Disson, O. & Lecuit, M. 2020. Maternal-neonatal listeriosis. *Virulence*, 11(1): 391–397. https://doi.org/10.1080/21505594.2020.1759287

Charlier, C., Kermorvant-Duchemin, E., Perrodeau, E., Moura, A., Maury, M.M., Bracq-Dieye, H., Thouvenot, P. *et al.* 2022. Neonatal listeriosis presentation and outcome: a prospective study of 189 cases. *Clinical Infectious Diseases,* 74(1): 8–16. https://doi.org/10.1093/cid/ciab337

Chen, Y., Ross, W.H., Scott, V.N. & Gombas, D.E. 2003. *Listeria monocytogenes*: low levels equal low risk. *Journal of Food Protection*, 66(4): 570–577. https://doi.org/10.4315/0362-028x-66.4.570

Chen, Y., Zhang, W. & Knabel, S.J. 2007. Multi-virulence-locus sequence typing identifies single nucleotide polymorphisms which differentiate epidemic clones and outbreak strains of *Listeria monocytogenes. Journal of Clinical Microbiology*, 45(3): 835–846. https://doi.org/10.1128/JCM.01575-06

Chen, Y., Ross, W.H., Whiting, R.C., Van Stelten, A., Nightingale, K.K., Wiedmann, M. & Scott, V.N. 2011. Variation in *Listeria monocytogenes* dose responses in relation to subtypes encoding a full-length or truncated internalin A. *Applied and Environmental Microbiology*, 77(4): 1171–1180. https://doi.org/10.1128/AEM.01564-10

Chen, Y., Dennis, S.B., Hartnett, E., Paoli, G., Pouillot, R., Ruthman, T. & Wilson, M. 2013. FDA-iRISK--a comparative risk assessment system for evaluating and ranking food-hazard pairs: case studies on microbial hazards. *Journal of Food Protection*, 76(3): 376–385. https://doi.org/10.4315/0362-028X.JFP-12-372

Chen, Y., Burall, L.S., Luo, Y., Timme, R., Melka, D., Muruvanda, T., Payne, J., Wang, C., Kastanis, G., Maounounen-Laasri, A., De Jesus, A.J., Curry, P.E., Stones, R., K'Aluoch, O., Liu, E., Salter, M., Hammack, T.S., Evans, P.S., Parish, M., Allard, M.W., Datta, A., Strain, E.A. & Brown, E.W. 2016. *Listeria monocytogenes* in stone fruits linked to a multistate outbreak: enumeration of cells and whole-genome sequencing. *Applied and Environmental Microbiology*, 82(24): 7030–7040. https://doi.org/10.1128/AEM.01486-16

Chen, Y., Luo, Y., Curry, P., Timme, R., Melka, D., Doyle, M., Parish, M., Hammack, T.S., Allard, M.W., Brown, E.W. & Strain, E.A. 2017. Assessing the genome level diversity of *Listeria monocytogenes* from contaminated ice cream and environmental samples linked to a listeriosis outbreak in the United States. *PloS One*, 12(2): e0171389. https://doi.org/10.1371/journal.pone.0171389

Chen, Y., Qian, C., Liu, C., Shen, H., Wang, Z., Ping, J., Wu, J. & Chen, H. 2020. Nucleic acid amplification free biosensors for pathogen detection. *Biosensors & Bioelectronics*, 153: 112049. https://doi.org/10.1016/j.bios.2020.112049

Chenal-Francisque, V., Lopez, J., Cantinelli, T., Caro, V., Tran, C., Leclercq, A., Lecuit, M. & Brisse, S. 2011. Worldwide distribution of major clones of *Listeria monocytogenes*. *Emerging Infectious Diseases*, 17(6): 1110–1112. https://doi.org/10.3201/eid1706.101778

Chenal-Francisque, V., Maury, M.M., Lavina, M., Touchon, M., Leclercq, A., Lecuit, M. & Brisse, S. 2015. Clonogrouping, a rapid multiplex PCR method for identification of major clones of *Listeria monocytogenes*. *Journal of Clinical Microbiology*, 53(10): 3355–3358. https://doi.org/10.1128/JCM.00738-15

Christiansen, J.K., Larsen, M.H., Ingmer, H., Søgaard-Andersen, L. & Kallipolitis, B.H. 2004. The RNA-binding protein Hfq of *Listeria monocytogenes*: role in stress tolerance and virulence. *Journal of Bacteriology*, 186(11): 3355–3362. https://doi.org/10.1128/JB.186.11.3355-3362.2004

Clayton, E.M., Daly, K.M., Guinane, C.M., Hill, C., Cotter, P.D. & Ross, P.R. 2014. Atypical *Listeria innocua* strains possess an intact LIPI-3. *BMC Microbiology*, 14: 58. https://doi.org/10.1186/1471-2180-14-58

Cordero, N., Maza, F., Navea-Perez, H., Aravena, A., Marquez-Fontt, B., Navarrete, P., Figueroa, G., González, M., Latorre, M. & Reyes-Jara, A. 2016. Different transcriptional responses from slow and fast growth rate strains of *Listeria monocytogenes* adapted to low temperature. *Frontiers in Microbiology*, 7: 229. https://doi.org/10.3389/fmicb.2016.00229

Cotter, P.D., Draper, L.A., Lawton, E.M., Daly, K.M., Groeger, D.S., Casey, P.G., Ross, R.P. & Hill, C. 2008. Listeriolysin S, a novel peptide haemolysin associated with a subset of lineage I *Listeria monocytogenes. PLoS Pathogens*, 4(9): e1000144. https://doi.org/10.1371/journal.ppat.1000144

Cox, L.J., Keller, N. & Van Schothorst, M. 1988. The use and misuse of quantitative determinations of *Enterobacteriaceae* in food microbiology. *Society for Applied Bacteriology Symposium Series*, 17: 237S-249S.

Cressey, P.J., Lake, R.J., Thornley, C. & Campbell, D. 2019. Expert elicitation for estimation of the proportion foodborne for selected microbial pathogens in New Zealand. *Foodborne Pathogens and Disease*, 16(8): 543–549. https://doi.org/10.1089/fpd.2018.2576

Currie, A., Farber, J.M., Nadon, C., Sharma, D., Whitfield, Y., Gaulin, C., Galanis, E., Bekal, S., Flint, J., Tschetter, L., Pagotto, F., Lee, B., Jamieson, F., Badiani, T., MacDonald, D., Ellis, A., May-Hadford, J., McCormick, R., Savelli, C., Middleton, D., Allen, V., Tremblay, F.-W., MacDougall, L., Hoang, L., Shyng, S., Everett, D., Chui, L., Louie, M., Bangura, H., Levett, P.N., Wilkinson, K., Wylie, J., Reid, J., Major, B., Engel, D., Douey, D., Huszczynski, G., Lecci, J.D., Strazds, J., Rousseau, J., Ma, K., Isaac, L. & Sierpinska, U. 2015. Multi-province listeriosis outbreak linked to contaminated Deli meat consumed primarily in institutional settings, Canada, 2008. *Foodborne Pathogens and Disease*, 12(8): 645–652. https://doi.org/10.1089/fpd.2015.1939

Dairy Authority of South Australia. 2015. *Microbiological testing criteria – Minimum testing requirements for manufacturers of dairy food products.* https://dairy-safe.com.au/wp-content/uploads/Dairysafe-Microbiological-Testing-Criteria-FOR-WEB.pdf

Dalton, C.B., Merritt, T.D., Unicomb, L.E., Kirk, M.D., Stafford, R.J., Lalor, K. & OzFoodNet Working Group. 2011. A national case-control study of risk factors for listeriosis in Australia. *Epidemiology and Infection*, 139(3): 437–445. https://doi.org/10.1017/S0950268810000944

Davidson, V.J., Ravel, A., Nguyen, T.N., Fazil, A. & Ruzante, J.M. 2011. Food-specific attribution of selected gastrointestinal illnesses: estimates from a Canadian expert elicitation survey. *Foodborne Pathogens and Disease*, 8(9): 983–995. https://doi.org/10.1089/fpd.2010.0786

de las Heras, A., Cain, R.J., Bielecka, M.K. & Vázquez-Boland, J.A. 2011. Regulation of *Listeria* virulence: PrfA master and commander. *Current Opinion in Microbiology*, 14(2): 118–127. https://doi.org/10.1016/j.mib.2011.01.005

den Besten, H., Wells-Bennik, M. & Zwietering, M.H. 2018. Natural diversity in heat resistance of bacteria and bacterial spores: impact on food safety and quality. *Annual Review of Food Science and Technology*, 9: 383–410. https://doi.org/10.1146/annurev-food-030117-012808

Desai, A.N., Anyoha, A., Madoff, L.C. & Lassmann, B. 2019. Changing epidemiology of *Listeria monocytogenes* outbreaks, sporadic cases, and recalls globally: A review of ProMED reports from 1996 to 2018. *International Journal of Infectious Diseases*, 84: 48–53. https://doi.org/10.1016/j.ijid.2019.04.021

Doumith, M., Buchrieser, C., Glaser, P., Jacquet, C. & Martin, P. 2004. Differentiation of the major *Listeria monocytogenes* serovars by multiplex PCR. *Journal of Clinical Microbiology*, 42(8): 3819–3822. https://doi.org/10.1128/JCM.42.8.3819-3822.2004

Drolia, R. & Bhunia, A. K. 2019. Crossing the intestinal barrier via *Listeria* adhesion protein and internalin A. *Trends in Microbiology*, 27(5): 408–425. https://doi.org/10.1016/j.tim.2018.12.007

Ducey, T.F., Page, B., Usgaard, T., Borucki, M.K., Pupedis, K. & Ward, T.J. 2007. A single-nucleotide-polymorphism-based multilocus genotyping assay for subtyping lineage I isolates of *Listeria monocytogenes*. *Applied and Environmental Microbiology*, 73(1): 133–147. https://doi.org/10.1128/AEM.01453-06

EFSA (European Food Safety Authority). 2014. Guidance on expert knowledge elicitation in food and feed safety risk assessment. *EFSA Journal*, 12(6): 3734. www.efsa.europa.eu/de/efsajournal/pub/3734

EFSA. 2017. Hazard analysis approaches for certain small retail establishments in view of the application of their food safety management systems. *EFSA Journal*, 15(3): 4697. https://efsa.onlinelibrary.wiley.com/doi/full/10.2903/j.efsa.2017.4697

EFSA. 2018. *Listeria monocytogenes* contamination of ready-to-eat foods and the risk for human health in the EU. EFSA BIOHAZ Panel (EFSA Panel on Biological Hazards). *EFSA Journal*, 16(1): 5134. https://doi.org/10.2903/j.efsa.2018.5134

EFSA. 2020. The public health risk posed by *Listeria monocytogenes* in frozen fruit and vegetables including herbs, blanched during processing. *EFSA Journal*, 18(4): 6092. https://efsa.onlinelibrary.wiley.com/doi/10.2903/j.efsa.2020.6092

EFSA (European Food Safety Authority) & ECDC (European Centre for Disease Prevention and Control). 2015. The European Union summary report on trends and sources of zoonoses, zoonotic agents and food-borne outbreaks in 2013. *EFSA Journal*, 13(1): 3991. https://doi.org/10.2903/j.efsa.2015.3991

EFSA & ECDC. 2017. The European Union summary report on trends and sources of zoonoses, zoonotic agents and food-borne outbreaks in 2016. *EFSA Journal*, 15(12): 5077. https://doi.org/10.2903/j.efsa.2017.5077.

EFSA & ECDC. 2018a. Multi-country outbreak of *Listeria monocytogenes* serogroup IVb, multi-locus sequence type 6, infections probably linked to frozen corn. *EFSA Supporting Publications,* EN-1402. https://doi.org/10.2903/sp.efsa.2018.EN-1402

EFSA & ECDC. 2018b. Multi-country outbreak of *Listeria monocytogenes* serogroup IVb, multi-locus sequence type 6, infections linked to frozen corn and possibly to other frozen vegetables – first update. *EFSA Supporting Publications,* EN-1448. https://doi.org/10.2903/sp.efsa.2018.EN-1448

EFSA & ECDC. 2019a. Multi-country outbreak of *Listeria monocytogenes* clonal complex 8 infections linked to consumption of cold-smoked fish products. *EFSA Supporting Publications,* EN-1665. https://doi.org/10.2903/sp.efsa.2019.EN-1665

EFSA & ECDC. 2019b. Multi-country outbreak of *Listeria monocytogenes* sequence type 6 infections linked to ready-to-eat meat products. *EFSA Supporting Publications*, EN-1745. https://doi.org/10.2903/sp.efsa.2019.EN-1745

Evans, E.W. & Redmond, E.C. 2014. Behavioral risk factors Associated with listeriosis in the home: a review of consumer food safety studies. *Journal of Food Protection,* 77(3): 510–521. https://doi.org/10.4315/0362-028X.JFP-13-238

Falk, L.E., Fader, K.A., Cui, D.S., Totton, S.C., Fazil, A.M., Lammerding, A.M. & Smith, B.A. 2016. Comparing listeriosis risks in at-risk populations using a user-friendly quantitative microbial risk assessment tool and epidemiological data. *Epidemiology and Infection*, 144(13): 2743–2758. https://doi.org/10.1017/S0950268816000327

Faralla, C., Rizzuto, G.A., Lowe, D.E., Kim, B., Cooke, C., Shiow, L.R. & Bakardjiev, A.I. 2016. InlP, a new virulence factor with strong placental tropism. *Infection and Immunity*, 84(12): 3584–3596. https://doi.org/10.1128/IAI.00625-16

Faralla, C., Bastounis, E.E., Ortega, F.E., Light, S.H., Rizzuto, G., Gao, L., Marciano, D.K., Nocadello, S., Anderson, W.F., Robbins, J.R., Theriot, J.A. & Bakardjiev, A.I. 2018. *Listeria monocytogenes* InlP interacts with afadin and facilitates basement membrane crossing. *PLoS Pathogens*, 14(5): e1007094. https://doi.org/10.1371/journal.ppat.1007094

Farber, J.M., Zwietering, M., Wiedmann, M., Schaffner, D., Hedberg, C.W., Harrison, M.A., Hartnett, E., Chapman, B., Donnelly, C.W., Goodburn, K.E. & Gummalla, S. 2021. Alternative approaches to the risk management of *Listeria monocytogenes* in low risk foods. *Food Control*, 123: 107601. https://doi.org/10.1016/j.foodcont.2020.107601

FAO (Food and Agriculture Organization of the United Nations). 1999. *FAO Expert consultation on the trade impact of Listeria in fish products.* FAO Fisheries report 604. Rome. www.fao.org/3/x3018e/x3018e00.htm

FAO. 2007. *International standards for phytosanitary measures, ISPM No. 5, Glossary of phytosanitary terms.* Rome. www.ippc.int/largefiles/adopted_ISPMs_previousversions/en/ISPM_05_2007_En_2007-07-26.pdf

FAO & WHO (World Health Organization). 1999. *Report of the thirty second session of the Codex Committee on Food Hygiene.* Rome, FAO. www.fao.org/fileadmin/templates/agns/pdf/jemra/SL00_en.pdf

FAO & WHO. 2000. *Report of the Joint FAO/WHO Expert Consultation on Risk Assessment of Microbiological Hazards in Foods.* Rome, FAO. www.fao.org/fao-who-codexalimentarius/sh-proxy/en/?lnk=1&url=https%253A%252F%252Fworkspace.fao.org%252Fsites%252Fcodex%252FMeetings%252FCX-712-32%252FAl01_13e.pdf

FAO & WHO. 2004a. *Risk assessment of Listeria monocytogenes in ready-to-eat foods: interpretive summary.* Microbiological Risk Assessment Series No. 4. Geneva, WHO. www.who.int/publications/i/item/risk-assessment-of-listeria-monocytogenes-in-ready-to-eat-foods-interpretive-summary

FAO & WHO. 2004b. *Risk assessment of Listeria monocytogenes in ready-to-eat foods: technical report.* Microbiological Risk Assessment Series No. 5. Geneva, WHO. www.who.int/publications/i/item/risk-assessment-of-listeria-monocytogenes-in-ready-to-eat-foods

FAO & WHO. 2009. *Codex Alimentarius. Guidelines on the application of general principles of food hygiene to the control of Listeria Monocytogenes in foods.* CAC/GL 61 - 2007. Rome, FAO. https://www.fao.org/fao-who-codexalimentarius/sh-proxy/en/?lnk=1&url=https%253A%252F%252Fworkspace.fao.org%252Fsites%252Fcodex%252FStandards%252FCXG%2B61-2007%252FCXG_061e.pdf

FAO & WHO. 2019. *Food control system assessment tool - Introduction and glossary.* Food Safety and Quality Series No. 7/1. Rome, FAO. https://www.fao.org/3/ca5334en/CA5334EN.pdf

FAO & WHO. 2020. *Codex Alimentarius. General principles of food hygiene.* CXC 1-1969. Rome, FAO. https://www.fao.org/fao-who-codexalimentarius/sh-proxy/en/?lnk=1&url=https%253A%252F%252Fworkspace.fao.org%252Fsites%252Fcodex%252FStandards%252FCXC%2B1-1969%252FCXC_001e.pdf

FDA (Food and Drug Administration) & FSIS (Food Safety Inspection Service) of the United States of America. 2003. Quantitative assessment of relative risk to public health from foodborne *Listeria monocytogenes* among selected categories of ready-to-eat foods. In: *FDA US Food and Drug Administration.* Washington, D.C. Cited 20 June 2021. www.fda.gov/food/cfsan-risk-safety-assessments/quantitative-assessment-relative-risk-public-health-foodborne-listeria-monocytogenes-among-selected

Félix, B., Feurer, C., Maillet, A., Guillier, L., Boscher, E., Kerouanton, A., Denis, M. & Roussel, S. 2018. Population genetic structure of *Listeria monocytogenes* strains isolated from the pig and pork production chain in France. *Frontiers in Microbiology*, 9: 684. https://doi.org/10.3389/fmicb.2018.00684

Ferguson. B. 2020. Trends in Food Safety Testing. *Trends in Food Safety,* 24 June 2020. Troy, MI, USA. Cited 5 July 2021. https://www.food-safety.com/articles/6666-trends-in-food-safety-testing

Fernàndez-Sabé, N., Cervera, C., López-Medrano, F., Llano, M., Sáez, E., Len, O., Fortún, J., Blanes, M., Laporta, R., Torre-Cisneros, J., Gavaldà, J., Muñoz, P., Fariñas, M.C., María Aguado, J., Moreno, A. & Carratalà, J. 2009. Risk factors, clinical features, and outcomes of listeriosis in solid-organ transplant recipients: a matched case-control study. *Clinical Infectious Diseases,* 49(8): 1153–1159. https://doi.org/10.1086/605637

Ferreira, V., Wiedmann, M., Teixeira, P. & Stasiewicz, M. J. 2014. *Listeria monocytogenes* persistence in food-associated environments: epidemiology, strain characteristics, and implications for public health. *Journal of Food Protection,* 77(1): 150–170. https://doi.org/10.4315/0362-028X.JFP-13-150

Filipello, V., Mughini-Gras, L., Gallina, S., Vitale, N., Mannelli, A., Pontello, M., Decastelli, L., Allard, M. W., Brown, E.W. & Lomonaco, S. 2020. Attribution of *Listeria monocytogenes* human infections to food and animal sources in Northern Italy. *Food Microbiology,* 89: 103433. https://doi.org/10.1016/j.fm.2020.103433

Foddai, A.C.G. & Grant, I.R. 2020. Methods for detection of viable foodborne pathogens: current state-of-art and future prospects. *Applied Microbiology and Biotechnology,* 104(10): 4281–4288. https://doi.org/10.1007/s00253-020-10542-x

Friesema, I.H., Kuiling, S., van der Ende, A., Heck, M.E., Spanjaard, L. & van Pelt, W. 2015. Risk factors for sporadic listeriosis in the Netherlands, 2008 to 2013. *Eurosurveillance,* 20(31): 21199. https://doi.org/10.2807/1560-7917.es2015.20.31.21199

Fritsch, L., Guillier, L. & Augustin, J.-C. 2018. Next generation quantitative microbiological risk assessment: Refinement of the cold smoked salmon-related listeriosis risk model by integrating genomic data. *Microbial Risk Analysis,* 10: 20–27. https://doi.org/10.1016/j.mran.2018.06.003

FSA (Food Standards Agency) & FSS (Food Standards Scotland). 2017. *Estimating quality adjusted life years and willingness to pay values for microbiological foodborne disease (Phase 2). Final Report.* London, Eftec. www.food.gov.uk/sites/default/files/media/document/fs102087p2finrep.pdf

FSIS (Food Safety and Inspection Service) & USDA (United States of America Department of Agriculture). 2014. *FSIS Compliance guideline: controlling Listeria monocytogenes in post-lethality exposed ready-to-eat meat and poultry products.* FSIS *Listeria* Guideline. www.mscbs.gob.es/profesionales/saludPublica/sanidadExterior/docs/Controlling-Lm-RTE-Guideline.pdf

FSIS & USDA. 2021. Isolation and identification of *Listeria monocytogenes* from red meat, poultry, ready-to-eat siluriformes (fish) and egg products, and environmental samples. In: *Microbiology Laboratory Guidebook*. Washington, DC. USDA. www.fsis.usda.gov/news-events/publications/microbiologylaboratory-guidebook

Gaballa, A., Guariglia-Oropeza, V., Wiedmann, M. & Boor, K.J. 2019. Cross talk between SigB and PrfA in *Listeria monocytogenes* facilitates transitions between extra- and intracellular environments. *Microbiology and Molecular Biology Reviews*, 83(4): e00034–19. https://doi.org/10.1128/MMBR.00034-19

Gasanov, U., Hughes, D. & Hansbro, P.M. 2005. Methods for the isolation and identification of *Listeria* spp. and *Listeria monocytogenes*: a review. *FEMS Microbiology Reviews*, 29(5): 851–875. https://doi.org/10.1016/j.femsre.2004.12.002

Gerner-Smidt, P., Ethelberg, S., Schiellerup, P., Christensen, J.J., Engberg, J., Fussing, V., Jensen, A., Jensen, C., Petersen, A.M. & Bruun, B.G. 2005. Invasive listeriosis in Denmark 1994-2003: a review of 299 cases with special emphasis on risk factors for mortality. *Clinical Microbiology and Infection*, 11(8): 618–624. https://doi.org/10.1111/j.1469-0691.2005.01171.x

Garner, D. & Kathariou, S. 2016. Fresh produce-associated listeriosis outbreaks, sources of concern, teachable moments, and insights. *Journal of Food Protection*, 79(2): 337–344. https://doi.org/10.4315/0362-028X.JFP-15-387

Gillesberg Lassen, S., Ethelberg, S., Björkman, J.T., Jensen, T., Sørensen, G., Kvistholm Jensen, A., Müller, L., Nielsen, E.M. & Mølbak, K. 2016. Two listeria outbreaks caused by smoked fish consumption-using whole-genome sequencing for outbreak investigations. *Clinical Microbiology and Infection,* 22(7): 620–624. https://doi.org/10.1016/j.cmi.2016.04.017

Gillespie, I.A., Mook, P., Little, C.L., Grant, K.A. & McLauchlin, J. 2010. Human listeriosis in England, 2001-2007: association with neighbourhood deprivation. *Eurosurveillance*, 15(27): 7–16. https://doi.org/10.2807/ese.15.27.19609-en

Gkogka, E., Reij, M.W., Havelaar, A.H., Zwietering, M.H. & Gorris, L.G. 2011. Risk-based estimate of effect of foodborne diseases on public health, Greece. *Emerging Infectious Diseases*, 17(9) : 1581–1590. https://doi.org/10.3201/eid1709.101766

Gnanou-Besse, N., Lombard, B., Guillier, L., François, D., Romero, K., Pierru, S., Bouhier, L. & Rollier, P. 2019. Validation of standard method EN ISO 11290 - Part 1 - Detection of *Listeria monocytogenes* in food. *International Journal of Food Microbiology*, 288: 13–21. https://doi.org/10.1016/j.ijfoodmicro.2018.03.024

Goulet, V., Hebert, M., Hedberg, C., Laurent, E., Vaillant, V., De Valk, H. & Desenclos, J. C. 2012a. Incidence of listeriosis and related mortality among groups at risk of acquiring listeriosis. *Clinical Infectious Diseases,* 54(5): 652–660. https://doi.org/10.1093/cid/cir902

Goulet, V., Leclercq, A., Laurent, E., King, L.A., Chenal-Francisque, V., Vaillant, V., Letort, M.J., Lecuit, M. & De Valk, H., 2012b. Surveillance de la listériose humaine en France, 1999-2011. *BEH Hors-Série*, 38–40.

Guo, Y., Zhao, C., Liu, Y., Nie, H., Guo, X., Song, X., Xu, K., Li, J. & Wang, J. 2020. A novel fluorescence method for the rapid and effective detection of *Listeria monocytogenes* using aptamer-conjugated magnetic nanoparticles and aggregation-induced emission dots. *The Analyst*, 145(11): 3857–3863. https://doi.org/10.1039/d0an00397b

Haagsma, J.A., Van der Zanden, B.P., Tariq, L., Van Pelt, W., Van Duynhoven, Y.T.H.P. & Havelaar, A.H. 2009. Disease burden and costs of selected foodborne pathogens in the Netherlands, 2006. *RIVM Rapport*, 330331001. https://www.rivm.nl/bibliotheek/rapporten/330331001.pdf

Haase, J.K., Didelot, X., Lecuit, M., Korkeala, H., L. monocytogenes MLST Study Group & Achtman, M. 2014. The ubiquitous nature of *Listeria monocytogenes* clones: a large-scale multilocus sequence typing study. *Environmental Microbiology*, 16(2): 405–416. https://doi.org/10.1111/1462-2920.12342

Halbedel, S., Wilking, H., Holzer, A., Kleta, S., Fischer, M.A., Lüth, S., Pietzka, A., Huhulescu, S., Lachmann, R., Krings, A., Ruppitsch, W., Leclercq, A., Kamphausen, R., Meincke, M., Wagner-Wiening, C., Contzen, M., Kraemer, I.B., Al Dahouk, S., Allerberger, F., Stark, K. & Flieger, A. 2020. Large nationwide outbreak of invasive listeriosis associated with blood sausage, Germany, 2018-2019. *Emerging Infectious Diseases*, 26(7): 1456–1464. https://doi.org/10.3201/eid2607.200225

Hamon, M., Bierne, H. & Cossart, P. 2006. *Listeria monocytogenes*: a multifaceted model. *Nature Reviews Microbiology*, 4(6): 423–434. https://doi.org/10.1038/nrmicro1413

Hansen, S., Hall, M., Grundström, C., Brännström, K., Sauer-Eriksson, A.E. & Johansson, J. 2020. A novel growth-based selection strategy identifies new constitutively active variants of the major virulence Regulator PrfA in *Listeria monocytogenes*. *Journal of Bacteriology*, 202(11): e00115–20. https://doi.org/10.1128/JB.00115-20

Harter, E., Wagner, E. M., Zaiser, A., Halecker, S., Wagner, M. & Rychli, K. 2017. Stress survival Islet 2, predominantly present in *Listeria monocytogenes* strains of sequence type 121, is involved in the alkaline and oxidative stress responses. *Applied and Environmental Microbiology*, 83(16): e00827–17. https://doi.org/10.1128/AEM.00827-17

Havelaar, A.H., Haagsma, J.A., Mangen, M.J., Kemmeren, J.M., Verhoef, L.P., Vijgen, S.M., Wilson, M., Friesema, I.H., Kortbeek, L.M., van Duynhoven, Y.T. & van Pelt, W. 2012. Disease burden of foodborne pathogens in the Netherlands, 2009. *International Journal of Food Microbiology*, 156(3): 231–238. https://doi.org/10.1016/j.ijfoodmicro.2012.03.029

Havelaar, A.H., Kirk, M.D., Torgerson, P.R., Gibb, H.J., Hald, T., Lake, R.J., Praet, N., Bellinger, D.C., de Silva, N.R., Gargouri, N., Speybroeck, N., Cawthorne, A., Mathers, C., Stein, C., Angulo, F.J., Devleesschauwer, B. & World Health Organization Foodborne Disease Burden Epidemiology Reference Group. 2015. World Health Organization global estimates and regional comparisons of the burden of foodborne disease in 2010. *PLoS Medicine*, 12(12): e1001923. https://doi.org/10.1371/journal.pmed.1001923

Health Canada. 2015. Laboratory procedures for the microbiological analysis of foods - compendium of analytical methods. In: *Canada.ca*. Ottowa, Ontario, Canada, Government of Canada. https://www.canada.ca/en/health-canada/services/food-nutrition/research-programs-analytical-methods/analytical-methods/compendiummethods/laboratory-procedures-microbiological-analysis-foods-compendiumanalytical-methods.html

Hedberg, C. 2006. *Listeria* in Europe: the need for a European surveillance network is growing. *Eurosurveillance*, 11(6): 75–76.

Hingston, P., Chen, J., Dhillon, B.K., Laing, C., Bertelli, C., Gannon, V., Tasara, T., Allen, K., Brinkman, F. S., Truelstrup Hansen, L. & Wang, S. 2017. Genotypes associated with *Listeria monocytogenes* isolates displaying impaired or enhanced tolerances to cold, salt, acid, or desiccation stress. *Frontiers in Microbiology*, 8: 369. https://doi.org/10.3389/fmicb.2017.00369

Hitchins, A.D., Jinneman, K. & Chen, Y. 2017. Bacteriological Analytical Manual (BAM): Detection and Enumeration of *Listeria monocytogenes*. In: *USFDA*. Washington, DC, USFDA. Cited on 22 April 2022. https://www.fda.gov/food/laboratory-methods-food/bam-chapter-10-detection-listeria-monocytogenes-foods-and-environmental-samples-and-enumeration

Hoelzer, K., Pouillot, R. & Dennis, S. 2012a. Animal models of listeriosis: a comparative review of the current state of the art and lessons learned. *Veterinary Research*, 43(1): 18. https://doi.org/10.1186/1297-9716-43-18

Hoelzer, K., Pouillot, R. & Dennis, S. 2012b. *Listeria monocytogenes* growth dynamics on produce: a review of the available data for predictive modeling. *Foodborne Pathogens and Disease*, 9(7): 661–673. https://doi.org/10.1089/fpd.2011.1087

Hoelzer, K., Chen, Y., Dennis, S., Evans, P., Pouillot, R., Silk, B. J. & Walls, I. 2013. New data, strategies, and insights for *Listeria monocytogenes* dose-response models: summary of an interagency workshop, 2011. *Risk Analysis*, 33(9): 1568–1581. https://doi.org/10.1111/risa.12005

Hoffmann, S., Batz, M.B. & Morris, J.G., Jr. 2012. Annual cost of illness and quality-adjusted life year losses in the United States due to 14 foodborne pathogens. *Journal of Food Protection*, 75(7): 1292–1302. https://doi.org/10.4315/0362-028X.JFP-11-417

Holcomb, D.L., Smith, M.A., Ware, G.O., Hung, Y.C., Brackett, R.E. & Doyle, M.P. 1999. Comparison of six dose-response models for use with food-borne pathogens. *Risk Analysis*, 19(6): 1091–1100. https://doi.org/10.1023/a:1007078527037

Houtsma, P.C., De Wit, J.C. & Rombouts, F.M. 1993. Minimum inhibitory concentration (MIC) of sodium lactate for pathogens and spoilage organisms occurring in meat products. *International Journal of Food Microbiology*, 20: 247–257.

Hurley, D., Luque-Sastre, L., Parker, C.T., Huynh, S., Eshwar, A.K., Nguyen, S.V., Andrews, N., Moura, A., Fox, E.M., Jordan, K., Lehner, A., Stephan, R. & Fanning, S. 2019. Whole-genome sequencing-based characterization of 100 *Listeria monocytogenes* isolates collected from food-processing environments over a four-year period. *mSphere*, 4(4): e00252–19. https://doi.org/10.1128/mSphere.00252-19

ICMSF (International Commission on Microbiological Specifications for Foods). 1996. *Microorganisms in Foods, Microbiological Specifications of Food Pathogens*. Vol. 5. London, Blackie Academic and Professional. 513 pp.

Imdad, A., Retzer, F., Thomas, L.S., McMillian, M., Garman, K., Rebeiro, P.F., Deppen, S.A., Dunn, J.R. & Woron, A. M. 2018. Impact of culture-independent diagnostic testing on recovery of enteric bacterial infections. *Clinical Infectious Diseases*, 66(12): 1892–1898. https://doi.org/10.1093/cid/cix1128

ISO (International Organization for Standardization). 2017a. Microbiology of the food chain — Horizontal method for the detection and enumeration of *Listeria monocytogenes* and of *Listeria* spp. — Part 1: Detection method (ISO standard 11290-1:2017).

ISO (International Organization for Standardization). 2017b. Microbiology of the food chain — Horizontal method for the detection and enumeration of *Listeria monocytogenes* and of *Listeria* spp. — Part 2: Enumeration method. (ISO standard 11290-2:2017).

Jackson, K.A., Gould, L.H., Hunter, J.C., Kucerova, Z. & Jackson, B. 2018. Listeriosis outbreaks associated with soft cheeses, United States, 1998-2014. *Emerging Infectious Diseases*, 24(6): 1116–1118. https://doi.org/10.3201/eid2406.171051

Jacquet, C., Doumith, M., Gordon, J.I., Martin, P.M., Cossart, P. & Lecuit, M. 2004. A molecular marker for evaluating the pathogenic potential of foodborne *Listeria monocytogenes*. *The Journal of Infectious Diseases*, 189(11): 2094–2100. https://doi.org/10.1086/420853

Jacquinet, S., Klamer, S., Leroy, M., Van Cauteren, D. & Devleesschauwer, B. 2018. *Voedsel- en watergerelateerde infectieziekten.* Epidemiologische surveillance in België, 2015 en 2016. Brussels, Wetenschappelijk Instituut Volksgezondheid (WIV-ISP). https://epidemio.wiv-isp.be/ID/reports/Voedsel-%20en%20watergerelateerde%20 infectieziekten%20-%20Epidemiologie%20-%20Jaarrapport%202015-2016.pdf

Jamali, H., Chai, L.C. & Thong, K.L. 2013. Detection and isolation of *Listeria* spp. and *Listeria monocytogenes* in ready-to-eat foods with various selective culture media. *Food Control*, 32(1): 19–24. https://doi.org/10.1016/j.foodcont.2012.11.033

Kemmeren, J.M., Mangen, M.J.J., Van Duynhoven, Y.T.H.P. & Havelaar, A.H. 2006. Priority setting of foodborne pathogens: disease burden and costs of selected enteric pathogens. *RIVM Rapport*, 330080001. www.rivm.nl/bibliotheek/rapporten/330080001.pdf

Kirk, M.D., Pires, S.M., Black, R.E., Caipo, M., Crump, J.A., Devleesschauwer, B., Döpfer, D., Fazil, A., Fischer-Walker, C.L., Hald, T., Hall, A. J., Keddy, K.H., Lake, R.J., Lanata, C.F., Torgerson, P.R., Havelaar, A.H. & Angulo, F.J. 2015. World Health Organization estimates of the global and regional disease burden of 22 foodborne bacterial, protozoal, and viral diseases, 2010: a data synthesis. *PLoS Medicine*, 12(12): e1001921. https://doi.org/10.1371/journal.pmed.1001921

Kropac, A.C., Eshwar, A.K., Stephan, R. & Tasara, T. 2019. New insights on the role of the pLMST6 plasmid in *Listeria monocytogenes* biocide tolerance and virulence. *Frontiers in Microbiology*, 10: 1538. https://doi.org/10.3389/fmicb.2019.01538

Kurpas, M., Osek, J., Moura, A., Leclercq, A., Lecuit, M. & Wieczorek, K. 2020. Genomic characterization of *Listeria monocytogenes* isolated from ready-to-eat meat and meat processing environments in Poland. *Frontiers in Microbiology*, 11: 1412. https://doi.org/10.3389/fmicb.2020.01412

Kvistholm Jensen, A., Simonsen, J. & Ethelberg, S. 2017. Use of proton pump inhibitors and the risk of listeriosis: a nationwide registry-based case-control study. *Clinical Infectious Diseases*, 64(7): 845–851. https://doi.org/10.1093/cid/ciw860

Kwong, J.C., Ratnasingham, S., Campitelli, M.A., Daneman, N., Deeks, S.L., Manuel, D.G., Allen, V.G., Bayoumi, A.M., Fazil, A., Fisman, D.N., Gershon, A.S., Gournis, E., Heathcote, E.J., Jamieson, F.B., Jha, P., Khan, K.M., Majowicz, S.E., Mazzulli, T., McGeer, A.J., Muller, M.P., Raut, A., Rea, E., Remis, R.S., Shahin, R., Wright, A.J., Zagorski, B. & Crowcroft, N.S. 2012. The impact of infection on population health: results of the Ontario burden of infectious diseases study. *PLoS One*, 7(9): e44103. https://doi.org/10.1371/journal.pone.0044103

Lado, B. & Yousef, A.E. 2007. Characteristics of *Listeria monocytogenes* important to food processors. Ch 6. In: E.T. Ryser and E.H. Marth, eds. *Listeria, Listeriosis and Food Safety*. 3rd ed. pp. 157–213. Boca Raton, Florida, USA, CRC Press Taylor & Francis Group.

Lahou, E. & Uyttendaele, M. 2014. Evaluation of three swabbing devices for detection of *Listeria monocytogenes* on different types of food contact surfaces. *International Journal of Environmental Research and Public Health*, 11(1): 804–814. https://doi.org/10.3390/ijerph110100804

Lake, R.J., Cressey, P.J., Campbell, D.M. & Oakley, E. 2010. Risk ranking for foodborne microbial hazards in New Zealand: burden of disease estimates. *Risk Analysis*, 30(5): 743–752. https://doi.org/10.1111/j.1539-6924.2009.01269.x

Las Heras, V., Clooney, A.G., Ryan, F.J., Cabrera-Rubio, R., Casey, P.G., Hueston, C.M., Pinheiro, J., Rudkin, J.K., Melgar, S., Cotter, P.D., Hill, C. & Gahan, C. 2019. Short-term consumption of a high-fat diet increases host susceptibility to *Listeria monocytogenes* infection. *Microbiome*, 7(1): 7. https://doi.org/10.1186/s40168-019-0621-x

Leclercq, A., Chenal-Francisque, V., Dieye, H., Cantinelli, T., Drali, R., Brisse, S. & Lecuit, M. 2011. Characterization of the novel *Listeria monocytogenes* PCR serogrouping profile IVb-v1. *International Journal of Food Microbiology*, 147(1): 74–77. https://doi.org/10.1016/j.ijfoodmicro.2011.03.010

Lecuit, M. 2020. *Listeria monocytogenes,* a model in infection biology. *Cellular Microbiology*, 22(4): e13186. https://doi.org/10.1111/cmi.13186

Ledlod, S., Bunroddith, K., Areekit, S., Santiwatanakul, S. & Chansiri, K. 2020. Development of a duplex lateral flow dipstick test for the detection and differentiation of *Listeria* spp. and *Listeria monocytogenes* in meat products based on loop-mediated isothermal amplification. *Journal of Chromatography B*, 1139: 121834. https://doi.org/10.1016/j.jchromb.2019.121834

Law, J.W., Ab Mutalib, N.S., Chan, K.G. & Lee, L.H. 2015. An insight into the isolation, enumeration, and molecular detection of *Listeria monocytogenes* in food. *Frontiers in Microbiology*, 6: 1227. https://doi.org/10.3389/fmicb.2015.01227

Li, J., Savagatrup, S., Nelson, Z., Yoshinaga, K. & Swager, T. M. 2020. Fluorescent Janus emulsions for biosensing of *Listeria monocytogenes*. *Proceedings of the National Academy of Sciences of the United States of America*, 117(22): 11923–11930. https://doi.org/10.1073/pnas.2002623117

Li, X.P., Wang, S.F., Hou, P.B., Liu, J., Du, P., Bai, L., Fanning, S., Zhang, H.N., Chen, Y.Z., Zhang, Y.K. & Kang, D.M. 2020. Nosocomial cross-infection of *hypervirulent Listeria monocytogenes* sequence type 87 in China. *Annals of Translational Medicine*, 8(9): 603. https://doi.org/10.21037/atm-19-2743

Little, C.L., Pires, S.M., Gillespie, I.A., Grant, K. & Nichols, G.L. 2010. Attribution of human *Listeria monocytogenes* infections in England and Wales to ready-to-eat food sources placed on the market: adaptation of the Hald *Salmonella* source attribution model. *Foodborne Pathogens and Disease*, 7(7): 749–756. https://doi.org/10.1089/fpd.2009.0439

Liu, Y., Orsi, R.H., Gaballa, A., Wiedmann, M., Boor, K.J. & Guariglia-Oropeza, V. 2019. Systematic review of the *Listeria monocytogenes* σ^B regulon supports a role in stress response, virulence and metabolism. *Future Microbiology*, 14: 801–828. https://doi.org/10.2217/fmb-2019-0072

Lokerse, R.F.A., Maslowska-Corker, K.A., van de Wardt, L.C. & Wijtzes, T. 2016. Growth capacity of *Listeria monocytogenes* in ingredients of ready-to-eat salads. *Food Control*, 60: 338–345. https://doi.org/10.1016/j.foodcont.2015.07.041

Maertens de Noordhout, C.M., Devleesschauwer, B., Angulo, F.J., Verbeke, G., Haagsma, J., Kirk, M., Havelaar, A. & Speybroeck, N. 2014. The global burden of listeriosis: a systematic review and meta-analysis. *The Lancet Infectious Diseases*, 14(11): 1073–1082. https://doi.org/10.1016/S1473-3099(14)70870-9

Maertens de Noordhout, C., Devleesschauwer, B., Haagsma, J.A., Havelaar, A.H., Bertrand, S., Vandenberg, O., Quoilin, S., Brandt, P.T. & Speybroeck, N. 2017. Burden of salmonellosis, campylobacteriosis and listeriosis: a time series analysis, Belgium, 2012 to 2020. *Eurosurveillance*, 22(38): 30615. https://doi.org/10.2807/1560-7917.ES.2017.22.38.30615

Maia, R.L., Teixeira, P. & Mateus, T.L. 2019. Risk communication strategies (on listeriosis) for high-risk groups. *Trends in Food Science & Technology*, 84: 68–70. https://doi.org/10.1016/j.tifs.2018.03.006

Mandin, P., Fsihi, H., Dussurget, O., Vergassola, M., Milohanic, E., Toledo-Arana, A., Lasa, I., Johansson, J. & Cossart, P. 2005. VirR, a response regulator critical for *Listeria monocytogenes* virulence. *Molecular Microbiology*, 57(5): 1367–1380. https://doi.org/10.1111/j.1365-2958.2005.04776.x

Mangen, M.J., Bouwknegt, M., Friesema, I.H., Haagsma, J.A., Kortbeek, L.M., Tariq, L., Wilson, M., van Pelt, W. & Havelaar, A.H. 2015. Cost-of-illness and disease burden of food-related pathogens in the Netherlands, 2011. *International Journal of Food Microbiology*, 196: 84–93. https://doi.org/10.1016/j.ijfoodmicro.2014.11.022

Marik, C.M., Zuchel, J., Schaffner, D.W. & Strawn, L.K. 2020. Growth and survival of *Listeria monocytogenes* on intact fruit and vegetable surfaces during postharvest handling: a systematic literature review. *Journal of Food Protection*, 83(1) : 108–128. https://doi.org/10.4315/0362-028X.JFP-19-283

Maury, M.M., Tsai, Y.H., Charlier, C., Touchon, M., Chenal-Francisque, V., Leclercq, A., Criscuolo, A., Gaultier, C., Roussel, S., Brisabois, A., Disson, O., Rocha, E., Brisse, S. & Lecuit, M. 2016. Uncovering *Listeria monocytogenes* hypervirulence by harnessing its biodiversity. *Nature Genetics*, 48(3): 308–313. https://doi.org/10.1038/ng.3501

Maury, M.M., Chenal-Francisque, V., Bracq-Dieye, H., Han, L., Leclercq, A., Vales, G., Moura, A., Gouin, E., Scortti, M., Disson, O., Vázquez-Boland, J.A. & Lecuit, M. 2017. Spontaneous loss of virulence in natural populations of *Listeria monocytogenes. Infection and Immunity*, 85(11): e00541–17. https://doi.org/10.1128/IAI.00541-17

Maury, M.M., Bracq-Dieye, H., Huang, L., Vales, G., Lavina, M., Thouvenot, P., Disson, O., Leclercq, A., Brisse, S. & Lecuit, M. 2019. Hypervirulent *Listeria monocytogenes* clones' adaption to mammalian gut accounts for their association with dairy products. *Nature Communications*, 10(1): 2488. https://doi.org/10.1038/s41467-019-10380-0

McCollum, J.T., Cronquist, A.B., Silk, B.J., Jackson, K.A., O'Connor, K.A., Cosgrove, S., Gossack, J.P., Parachini, S.S., Jain, N.S., Ettestad, P., Ibraheem, M., Cantu, V., Joshi, M., DuVernoy, T., Fogg, N.W., Jr, Gorny, J.R., Mogen, K.M., Spires, C., Teitell, P., Joseph, L.A., Tarr, C.L., Imanishi, M., Neil, K.P., Tauxe, R.V. & Mahon, B.E. 2013. Multistate outbreak of listeriosis associated with cantaloupe. *The New England Journal of Medicine*, 369(10): 944–953. https://doi.org/10.1056/NEJMoa1215837

Meile, S., Sarbach, A., Du, J., Schuppler, M., Saez, C., Loessner, M.J. & Kilcher, S. 2020. Engineered reporter phages for rapid bioluminescence-based detection and differentiation of viable *Listeria* cells. *Applied and Environmental Microbiology*, 86(11): e00442–20. https://doi.org/10.1128/AEM.00442-20

Mejlholm, O., Gunvig, A., Borggaard, C., Blom-Hanssen, J., Mellefont, L., Ross, T., Leroi, F., Else, T., Visser, D. & Dalgaard, P. 2010. Predicting growth rates and growth boundary of *Listeria monocytogenes* - An international validation study with focus on processed and ready-to-eat meat and seafood. *International Journal of Food Microbiology*, 141(3): 137–150. https://doi.org/10.1016/j.ijfoodmicro.2010.04.026

Metselaar, K.I., den Besten, H.M., Boekhorst, J., van Hijum, S.A., Zwietering, M.H. & Abee, T. 2015. Diversity of acid stress resistant variants of *Listeria monocytogenes* and the potential role of ribosomal protein S21 encoded by rpsU. *Frontiers in Microbiology*, 6: 422. https://doi.org/10.3389/fmicb.2015.00422

Ministry of Health of People's Republic of China. 2016. *National standard of the People's Republic of China national of food safety standard, food microbiological examination: Listeria monocytogenes.* https://sppt.cfsa.net.cn:8086/staticPages/BC60573F-7E60-4E3F-AC6C-19E2484D73CD.html

Mook, P., Patel, B. & Gillespie, I. A. 2012. Risk factors for mortality in non-pregnancy-related listeriosis. *Epidemiology and Infection*, 140(4): 706–715. https://doi.org/10.1017/S0950268811001051

Morgand, M., Leclercq, A., Maury, M.M., Bracq-Dieye, H., Thouvenot, P., Vales, G., Lecuit, M. & Charlier, C. 2018. *Listeria monocytogenes*-associated respiratory infections: a study of 38 consecutive cases. *Clinical Microbiology and Infection*, 24(12): 1339.e1–1339.e5. https://doi.org/10.1016/j.cmi.2018.03.003

Moura, A., Criscuolo, A., Pouseele, H., Maury, M.M., Leclercq, A., Tarr, C., Björkman, J. T., Dallman, T., Reimer, A., Enouf, V., Larsonneur, E., Carleton, H., Bracq-Dieye, H., Katz, L. S., Jones, L., Touchon, M., Tourdjman, M., Walker, M., Stroika, S., Cantinelli, T., Chenal-Francisque, V., Kucerova, Z., Rocha, E.P.C., Nadon, C., Grant, K., Nielsen, E.M., Pot, B., Smidt, P.G., Lecuit, M. & Brisse, S. 2016. Whole genome-based population biology and epidemiological surveillance of *Listeria monocytogenes*. *Nature Microbiology*, 2: 16185. https://doi.org/10.1038/nmicrobiol.2016.185

Mughini-Gras, L., Kooh, P., Fravalo, P., Augustin, J.C., Guillier, L., David, J., Thébault, A., Carlin, F., Leclercq, A., Jourdan-Da-Silva, N., Pavio, N., Villena, I., Sanaa, M. & Watier, L. 2019. Critical orientation in the jungle of currently available methods and types of data for source attribution of foodborne diseases. *Frontiers in Microbiology*, 10: 2578. https://doi.org/10.3389/fmicb.2019.02578

Muhterem-Uyar, M., Dalmasso, M., Bolocan, A.S., Hernandez, M., Kapetanakou, A.E., Kuchta, T., Manios, S.G., Melero, B., Minarovičová, J., Nicolau, A.I, Rovira, J., Skandamis, P.N., Jordan, K., Rodríguez-Lázaro, D., Stessl, B. & Wagner, M. 2015. Environmental sampling for *Listeria monocytogenes* control in food-processing facilities reveals three contamination scenarios. *Food Control*, 51: 94–107. https://doi.org/10.1016/j.foodcont.2014.10.042

Newsome, R., Tran, N., Paoli, G.M., Jaykus, L.A., Tompkin, B., Miliotis, M., Ruthman, T., Hartnett, E., Busta, F.F., Petersen, B., Shank, F., McEntire, J., Hotchkiss, J., Wagner, M. & Schaffner, D.W. 2009. Development of a risk-ranking framework to evaluate potential high-threat microorganisms, toxins, and chemicals in food. *Journal of Food Science*, 74(2): R39-R45. https://doi.org/10.1111/j.1750-3841.2008.01042.x

Newsome, R., Balestrini, C.G., Baum, M.D., Corby, J., Fisher, W., Goodburn, K., Labuza, T.P., Prince, G., Thesmar, H.S. & Yiannas, F. 2014. Applications and perceptions of date labeling of food. *Comprehensive Reviews in Food Science and Food Safety*, 13(4): 745–769. https://doi.org/10.1111/1541-4337.12086

Nightingale, K.K., Windham, K. & Wiedmann, M. 2005. Evolution and molecular phylogeny of *Listeria monocytogenes* isolated from human and animal listeriosis cases and foods. *Journal of Bacteriology*, 187(16): 5537–5551. https://doi.org/10.1128/JB.187.16.5537-5551.2005

Nightingale, K.K., Ivy, R.A., Ho, A.J., Fortes, E.D., Njaa, B.L., Peters, R.M. & Wiedmann, M. 2008. inlA premature stop codons are common among *Listeria monocytogenes* isolates from foods and yield virulence-attenuated strains that confer protection against fully virulent strains. *Applied and Environmental Microbiology*, 74(21): 6570–6583. https://doi.org/10.1128/AEM.00997-08

Njage, P., Leekitcharoenphon, P., Hansen, L.T., Hendriksen, R.S., Faes, C., Aerts, M. & Hald, T. 2020. Quantitative microbial risk assessment based on whole genome sequencing data: case of *Listeria monocytogenes*. *Microorganisms*, 8(11): 1772. https://doi.org/10.3390/microorganisms8111772

Nouws, S., Bogaerts, B., Verhaegen, B., Denayer, S., Crombé, F., De Rauw, K., Piérard, D., Marchal, K., Vanneste, K., Roosens, N.H.C. & De Keersmaecker, S.C.J. 2020. The benefits of whole genome sequencing for foodborne outbreak investigation from the perspective of a national reference laboratory in a smaller country. *Foods*, 9(8): 1030. https://doi.org/10.3390/foods9081030

Oladimeji, V., Jeong, S., Almenar, E., Marks, B.P., Vorst, K.L., Brown, W. & Ryser, E.T. 2019. Predicting the growth of *Listeria monocytogenes* and *Salmonella* Typhimurium growth in diced celery, onions, and tomatoes during simulated commercial transport, retail storage, and display. *Journal of Food Protection*, 82: 287–300. https://doi.org/10.4315/0362-028X.JFP-18-277

Oliver, H.F., Orsi, R.H., Wiedmann, M. & Boor, K.J. 2010. *Listeria monocytogenes* σB has a small core regulon and a conserved role in virulence but makes differential contributions to stress tolerance across a diverse collection of strains. *Applied and Environmental Microbiology*, 76(13): 4216–4232. https://doi.org/10.1128/AEM.00031-10

Ooi, S.T. & Lorber, B. 2005. Gastroenteritis due to *Listeria monocytogenes*. *Clinical Infectious*, 40(9): 1327–1332. https://doi.org/10.1086/429324

Orsi, R.H., den Bakker, H.C. & Wiedmann, M. 2011. *Listeria monocytogenes* lineages: Genomics, evolution, ecology, and phenotypic characteristics. *International Journal of Medical Microbiology*, 301(2): 79–96. https://doi.org/10.1016/j.ijmm.2010.05.002

Østergaard, N.B., Eklöw, A. & Dalgaard, P. 2014. Modelling the effect of lactic acid bacteria from starter- and aroma culture on growth of *Listeria monocytogenes* in cottage cheese. *International Journal of Food Microbiology*, 188: 15–25. https://doi.org/10.1016/j.ijfoodmicro.2014.07.012

Ottesen, A., Ramachandran, P., Chen, Y., Brown, E., Reed, E. & Strain, E. 2020. Quasimetagenomic source tracking of *Listeria monocytogenes* from naturally contaminated ice cream. *BMC Infectious Diseases*, 20(1): 83. https://doi.org/10.1186/s12879-019-4747-z

Painter, J. & Slutsker, L. 2007. Listeriosis in humans. In: E.T. Ryser and E.H. Marth, eds. *Listeria, Listeriosis and Food Safety*. 3rd ed., pp. 85-111. Boca Raton, Florida, USA, CRC Press Taylor & Francis Group.

Pérez-Trallero, E., Zigorraga, C., Artieda, J., Alkorta, M. & Marimón, J.M. 2014. Two outbreaks of *Listeria monocytogenes* infection, Northern Spain. *Emerging Infectious Diseases*, 20(12): 2155–2157. https://doi.org/10.3201/eid2012.140993

Pietzka, A., Allerberger, F., Murer, A., Lennkh, A., Stöger, A., Cabal Rosel, A., Huhulescu, S., Maritschnik, S., Springer, B., Lepuschitz, S., Ruppitsch, W. & Schmid, D. 2019. Whole genome sequencing based surveillance of *L. monocytogenes* for early detection and investigations of listeriosis outbreaks. *Frontiers in Public Health*, 7: 139. https://doi.org/10.3389/fpubh.2019.00139

Pires, S.M., Evers, E.G., van Pelt, W., Ayers, T., Scallan, E., Angulo, F.J., Havelaar, A. & Hald, T. 2009. Attributing the human disease burden of foodborne infections to specific sources. *Foodborne Pathogens and Disease*, 6(4): 417–424. https://doi.org/10.1089/fpd.2008.0208

Pires, M.S., Jakobsen, L.S., Ellis-Iversen, J., Pessoa, J. & Ethelberg, S. 2020. Burden of disease estimates of seven pathogens commonly transmitted through foods in Denmark, 2017. *Foodborne Pathogens and Disease*, 17(5): 322–339. https://doi.org/10.1089/fpd.2019.2705

Pizarro-Cerdá, J., Kühbacher, A. & Cossart, P. 2012. Entry of *Listeria monocytogenes* in mammalian epithelial cells: an updated view. *Cold Spring Harbor Perspectives in Medicine*, 2(11): a010009. https://doi.org/10.1101/cshperspect.a010009

Pouillot, R., Hoelzer, K., Jackson, K.A., Henao, O.L. & Silk, B.J. 2012. Relative risk of listeriosis in Foodborne Diseases Active Surveillance Network (FoodNet) sites according to age, pregnancy, and ethnicity. *Clinical Infectious Diseases*, 54 Suppl 5: S405–S410. https://doi.org/10.1093/cid/cis269

Pouillot, R., Hoelzer, K., Chen, Y. & Dennis, S.B. 2015. *Listeria monocytogenes* dose response revisited--incorporating adjustments for variability in strain virulence and host susceptibility. *Risk Analysis*, 35(1): 90–108. https://doi.org/10.1111/risa.12235

Preussel, K., Milde-Busch, A., Schmich, P., Wetzstein, M., Stark, K. & Werber, D., 2016. Risk factors for sporadic non-pregnancy associated listeriosis in Germany-immunocompromised patients and frequently consumed ready-to-eat products. *PLoS One*, 10(11): e0142986.

Quereda, J.J., Dussurget, O., Nahori, M.A., Ghozlane, A., Volant, S., Dillies, M.A., Regnault, B., Kennedy, S., Mondot, S., Villoing, B., Cossart, P. & Pizarro-Cerda, J. 2016. Bacteriocin from epidemic *Listeria* strains alters the host intestinal microbiota to favor infection. *Proceedings of the National Academy of Sciences of the United States of America*, 113(20): 5706–5711. https://doi.org/10.1073/pnas.1523899113

Quereda, J.J., Meza-Torres, J., Cossart, P. & Pizarro-Cerdá, J. 2017. Listeriolysin S: A bacteriocin from epidemic *Listeria monocytogenes* strains that targets the gut microbiota. *Gut Microbes*, 8(4): 384–391. https://doi.org/10.1080/19490976.2017. 1290759

Quereda, J.J., Rodríguez-Gómez, I.M., Meza-Torres, J., Gómez-Laguna, J., Nahori, M.A., Dussurget, O., Carrasco, L., Cossart, P. & Pizarro-Cerdá, J. 2019. Reassessing the role of internalin B in *Listeria monocytogenes* virulence using the epidemic strain F2365. *Clinical Microbiology and Infection*, 25(2): 252.e1-252.e4. https://doi.org/10.1016/j.cmi.2018.08.022

Radomski, Nicolas. 2020. *La génomique bactérienne en appui à la santé publique - The bacterial genomics in support of public health.*

Radoshevich, L. & Cossart, P. 2018. *Listeria monocytogenes*: towards a complete picture of its physiology and pathogenesis. *Nature Reviews Microbiology*, 16(1): 32–46. https://doi.org/10.1038/nrmicro.2017.126

Rahman, A., Munther, D., Fazil, A., Smith, B. & Wu, J. 2018. Advancing risk assessment: mechanistic dose-response modelling of *Listeria monocytogenes* infection in human populations. *Royal Society Open Science*, 5(8): 180343. https://doi.org/10.1098/rsos.180343

Ribot, E.M., Freeman, M., Hise, K.B. & Gerner-Smidt, P. 2019. PulseNet: entering the age of next-generation sequencing. *Foodborne Pathogens and Disease*, 16(7): 451–456. https://doi.org/10.1089/fpd.2019.2634

Roche, S.M., Grépinet, O., Corde, Y., Teixeira, A.P., Kerouanton, A., Témoin, S., Mereghetti, L., Brisabois, A. & Velge, P. 2009. A *Listeria monocytogenes* strain is still virulent despite nonfunctional major virulence genes. *The Journal of Infectious Diseases*, 200(12): 1944–1948. https://doi.org/10.1086/648402

Rocourt, J. 1996. Risk factors for listeriosis. *Food Control*, 7: 195–202.

Rocourt, J., BenEmbarek, P., Toyofuku, H. & Schlundt, J. 2003. Quantitative risk assessment of *Listeria monocytogenes* in ready-to-eat foods: the FAO/WHO approach. *FEMS Immunology and Medical Microbiology*, 35(3): 263–267. https://doi.org/10.1016/S0928-8244(02)00468-6

Ronholm, J., Nasheri, N., Petronella, N. & Pagotto, F. 2016. Navigating microbiological food safety in the era of whole-genome sequencing. *Clinical Microbiology Reviews*, 29(4): 837–857. https://doi.org/10.1128/CMR.00056-16

Ross, T., Dalgaard, P. & Tienungoon, S. 2000. Predictive modelling of the growth and survival of *Listeria* in fishery products. *International Journal of Food Microbiology*, 62(3): 231–245. https://doi.org/10.1016/s0168-1605(00)00340-8

Ruzante, J.M., Davidson, V.J., Caswell, J., Fazil, A., Cranfield, J.A., Henson, S.J., Anders, S.M., Schmidt, C. & Farber, J.M. 2010. A multifactorial risk prioritization framework for foodborne pathogens. *Risk Analysis*, 30(5): 724–742. https://doi.org/10.1111/j.1539-6924.2009.01278.x

Ryan, S., Begley, M., Hill, C. & Gahan, C.G. 2010. A five-gene stress survival islet (SSI-1) that contributes to the growth of *Listeria monocytogenes* in suboptimal conditions. *Journal of Applied Microbiology*, 109(3): 984–995. https://doi.org/10.1111/j.1365-2672.2010.04726.x

Ryser, E.T. & Marth, E.H., eds. 1991. *Listeria, listeriosis, and food safety*. Marcel Dekker, New York, NY. 632 pp.

Scallan, E., Hoekstra, R.M., Mahon, B.E., Jones, T.F. & Griffin, P.M. 2015. An assessment of the human health impact of seven leading foodborne pathogens in the United States using disability adjusted life years. *Epidemiology and Infection*, 143(13): 2795–2804. https://doi.org/10.1017/S0950268814003185

Schardt, J., Jones, G., Müller-Herbst, S., Schauer, K., D'Orazio, S. & Fuchs, T.M. 2017. Comparison between *Listeria* sensu stricto and *Listeria* sensu lato strains identifies novel determinants involved in infection. *Scientific Reports*, 7(1): 17821. https://doi.org/10.1038/s41598-017-17570-0

Schjørring, S., Gillesberg Lassen, S., Jensen, T., Moura, A., Kjeldgaard, J.S., Müller, L., Thielke, S., Leclercq, A., Maury, M.M., Tourdjman, M., Donguy, M.P., Lecuit, M., Ethelberg, S. & Nielsen, E.M. 2017. Cross-border outbreak of listeriosis caused by cold-smoked salmon, revealed by integrated surveillance and whole genome sequencing (WGS), Denmark and France, 2015 to 2017. *Eurosurveillance*, 22(50): 17–00762. https://doi.org/10.2807/1560-7917.ES.2017.22.50.17-00762

Schlech, W.F., 3rd, Lavigne, P.M., Bortolussi, R.A., Allen, A.C., Haldane, E.V., Wort, A.J., Hightower, A.W., Johnson, S.E., King, S.H., Nicholls, E.S. & Broome, C.V. 1983. Epidemic listeriosis-evidence for transmission by food. *The New England Journal of Medicine*, 308(4): 203–206. https://doi.org/10.1056/NEJM198301273080407

Scobie, A., Kanagarajah, S., Harris, R.J., Byrne, L., Amar, C., Grant, K. & Godbole, G. 2019. Mortality risk factors for listeriosis - A 10 year review of non-pregnancy associated cases in England 2006-2015. *The Journal of Infection*, 78(3): 208–214. https://doi.org/10.1016/j.jinf.2018.11.007

Shen, A. & Higgins, D.E. 2006. The MogR transcriptional repressor regulates nonhierarchal expression of flagellar motility genes and virulence in *Listeria monocytogenes*. *PLoS Pathogens*, 2(4): e30. https://doi.org/10.1371/journal.ppat.0020030

Shinomiya, N., Tsuru, S., Fujisawa, H., Taniguchi, M., Zinnaka, Y. & Nomoto, K. 1988. Effect of a high-fat diet on resistance to *Listeria monocytogenes*. *Journal of Clinical & Laboratory Immunology*, 25(2): 97–100.

Silva, N.F.D., Neves, M.M.P.S., Magalhães, J.M.C.S., Freire, C. & Delerue-Matos, C. 2020. Emerging electrochemical biosensing approaches for detection of *Listeria monocytogenes* in food samples: An overview. *Trends in Food Science and Technology*, 99: 621–633. https://doi.org/10.1016/j.tifs.2020.03.031

Smith, A.M., Tau, N.P., Smouse, S.L., Allam, M., Ismail, A., Ramalwa, N.R., Disenyeng, B., Ngomane, M. & Thomas, J. 2019. Outbreak of *Listeria monocytogenes* in South Africa, 2017-2018: laboratory activities and experiences associated with whole-genome sequencing analysis of isolates. *Foodborne Pathogens and Disease*, 16(7): 524–530. https://doi.org/10.1089/fpd.2018.2586

Stoller, A., Stevens, M., Stephan, R. & Guldimann, C. 2019. Characteristics of *Listeria monocytogenes* strains persisting in a meat processing facility over a 4-year period. *Pathogens*, 8(1): 32. https://doi.org/10.3390/pathogens8010032

Tamburro, M., Ripabelli, G., Fanelli, I., Grasso, G.M. & Sammarco, M.L. 2010. Typing of *Listeria monocytogenes* strains isolated in Italy by inlA gene characterization and evaluation of a new cost-effective approach to antisera selection for serotyping. *Journal of Applied Microbiology*, 108(5): 1602–1611. https://doi.org/10.1111/j.1365-2672.2009.04555.x

Thomas, M.K., Vriezen, R., Farber, J.M., Currie, A., Schlech, W. & Fazil, A. 2015. Economic cost of a *Listeria monocytogenes* outbreak in Canada, 2008. *Foodborne Pathogens and Disease*, 12(12): 966–971. https://doi.org/10.1089/fpd.2015.1965

Thomas, J., Govender, N., McCarthy, K.M., Erasmus, L.K., Doyle, T.J., Allam, M., Ismail, A., Ramalwa, N., Sekwadi, P., Ntshoe, G., Shonhiwa, A., Essel, V., Tau, N., Smouse, S., Ngomane, H.M., Disenyeng, B., Page, N.A., Govender, N.P., Duse, A.G., Stewart, R., Thomas, T., Mahoney, D., Tourdjman, M., Disson, O., Thouvenot, P., Maury, M.M., Leclercq, A., Lecuit, M., Smith, A.M. & Blumberg, L.H. 2020. Outbreak of listeriosis in South Africa associated with processed meat. *The New England Journal of Medicine*, 382(7): 632–643. https://doi.org/10.1056/NEJMoa1907462

Tiensuu, T., Guerreiro, D.N., Oliveira, A.H., O'Byrne, C. & Johansson, J. 2019. Flick of a switch: regulatory mechanisms allowing *Listeria monocytogenes* to transition from a saprophyte to a killer. *Microbiology*, 165(8): 819–833. https://doi.org/10.1099/mic.0.000808

Tienungoon, S. 1998. Some aspects of the ecology of *Listeria monocytogenes* in salmonid aquaculture. Hobart, Tasmania, Australia, University of Tasmania. PhD Dissertation. https://eprints.utas.edu.au/22028/1/whole_TienungoonSuwunna1999_thesis.pdf

3M. 2021. Testing for indicator organisms and pathogens. In: *3M Food Safety News*. St. Paul, MN, USA. Cited 5 July 2021. https://food-safety-news.3m.com/fsn/improving-food-safety-by-testing-for-indicator-organisms-and-pathogens/

Välimaa, A.L., Tilsala-Timisjärvi, A. & Virtanen, E. 2015. Rapid detection and identification methods for *Listeria monocytogenes* in the food chain - A review. *Food Control*, 55: 103–114. https://doi.org/10.1016/j.foodcont.2015.02.037

Van Boeijen, I.K., Chavaroche, A.A., Valderrama, W.B., Moezelaar, R., Zwietering, M.H. & Abee, T. 2010. Population diversity of *Listeria monocytogenes* LO28: phenotypic and genotypic characterization of variants resistant to high hydrostatic pressure. *Applied and Environmental Microbiology*, 76(7): 2225–2233. https://doi.org/10.1128/AEM.02434-09

van Lier, A., McDonald, S.A., Bouwknegt, M., EPI group, Kretzschmar, M.E., Havelaar, A.H., Mangen, M.J., Wallinga, J. & de Melker, H. E. 2016. Disease burden of 32 infectious diseases in the Netherlands, 2007-2011. *PloS One*, 11(4): e0153106. https://doi.org/10.1371/journal.pone.0153106

Van Stelten, A., Simpson, J.M., Ward, T.J. & Nightingale, K.K. 2010. Revelation by single-nucleotide polymorphism genotyping that mutations leading to a premature stop codon in inlA are common among *Listeria monocytogenes* isolates from ready-to-eat foods but not human listeriosis cases. *Applied and Environmental Microbiology*, 76(9): 2783–2790. https://doi.org/10.1128/AEM.02651-09

Vázquez-Boland, J.A., Kuhn, M., Berche, P., Chakraborty, T., Domínguez-Bernal, G., Goebel, W., González-Zorn, B., Wehland, J. & Kreft, J. 2001. *Listeria* pathogenesis and molecular virulence determinants. *Clinical Microbiology Reviews*, 14(3): 584–640. https://doi.org/10.1128/CMR.14.3.584-640.2001

Velge, P., Herler, M., Johansson, J., Roche, S.M., Témoin, S., Fedorov, A. A., Gracieux, P., Almo, S.C., Goebel, W. & Cossart, P. 2007. A naturally occurring mutation K220T in the pleiotropic activator PrfA of *Listeria monocytogenes* results in a loss of virulence due to decreasing DNA-binding affinity. *Microbiology*, 153(Pt 4): 995–1005. https://doi.org/10.1099/mic.0.2006/002238-0

Wambogo, E.A., Vaudin, A.M., Moshfegh, A.J., Spungen, J.H., Doren, J. & Sahyoun, N.R. 2020. Toward a better understanding of listeriosis risk among older adults in the United States: characterizing dietary patterns and the sociodemographic and economic attributes of consumers with these patterns. *Journal of Food Protection*, 83(7): 1208–1217. https://doi.org/10.4315/JFP-19-617

Wambui, J., Eshwar, A.K., Aalto-Araneda, M., Pöntinen, A., Stevens, M., Njage, P. & Tasara, T. 2020. The analysis of field strains isolated from food, animal and clinical sources uncovers natural mutations in *Listeria monocytogenes* nisin resistance genes. *Frontiers in Microbiology*, 11: 549531. https://doi.org/10.3389/fmicb.2020.549531

Wang, L., Zhao, P., Si, X., Li, J., Dai, X., Zhang, K., Gao, S. & Dong, J. 2020. Rapid and specific detection of *Listeria monocytogenes* with an isothermal amplification and lateral flow strip combined method that eliminates false-positive signals from primer–dimers. *Frontiers in Microbiology*, 10: 2959. https://doi.org/10.3389/fmicb.2019.02959

Werber, D., Hille, K., Frank, C., Dehnert, M., Altmann, D., Müller-Nordhorn, J., Koch, J. & Stark, K. 2013. Years of potential life lost for six major enteric pathogens, Germany, 2004-2008. *Epidemiology and Infection*, 141(5): 961–968. https://doi.org/10.1017/S0950268812001550

Yang, Q., Xu, H., Zhang, Y., Liu, Y., Lu, X., Feng, X., Tan, J., Zhang, S. & Zhang, W. 2020. Single primer isothermal amplification coupled with SYBR Green II: Real-time and rapid visual method for detection of *Listeria monocytogenes* in raw chicken. LWT, 128: 109453. https://doi.org/10.1016/j.lwt.2020.109453

Yeni, F., Acar, S., Soyer, Y. & Alpas, H. 2017. How can we improve foodborne disease surveillance systems: A comparison through EU and US systems, *Food Reviews International*, 33(4): 406–423. https://www.semanticscholar.org/paper/How-can-we-improve-foodborne-disease-surveillance-A-Yeni-Acar/58f4a4db87cadb0002497b2e5f21d727287d278a

Yin, Y., Yao, H., Doijad, S., Kong, S., Shen, Y., Cai, X., Tan, W., Wang, Y., Feng, Y., Ling, Z., Wang, G., Hu, Y., Lian, K., Sun, X., Liu, Y., Wang, C., Jiao, K., Liu, G., Song, R., Chen, X., Pan, Z., Loessner, M.J. Chakraborty, T. & Jiao, X. 2019. A hybrid sub-lineage of *Listeria monocytogenes* comprising hypervirulent isolates. *Nature Communications*, 10(1): 4283. https://doi.org/10.1038/s41467-019-12072-1

Zeng, W., Vorst, K., Brown, W., Marks, B.P., Jeong, S., Pérez-Rodríguez, F. & Ryser, E.T. 2014. Growth of *Escherichia coli* O157:H7 and *Listeria monocytogenes* in packaged fresh-cut romaine mix at fluctuating temperatures during commercial transport, retail storage, and display. *Journal of Food Protection*, 77(2): 197–206. https://doi.org/10.4315/0362-028X.JFP-13-117

Ziegler, M., Kent, D., Stephan, R. & Guldimann, C. 2019. Growth potential of *Listeria monocytogenes* in twelve different types of RTE salads: Impact of food matrix, storage temperature and storage time. *International Journal of Food Microbiology*, 296: 83–92. https://doi.org/10.1016/j.ijfoodmicro.2019.01.016

Zwietering, M.H., Jacxsens., L, Membré, J.-M., Nauta, M. & Peterz, M. 2016. Relevance of microbial finished product testing in food safety management. *Food Control*, 60: 31–43, https://doi.org/10.1016/j.foodcont.2015.07.002

Annexes

Annex 1

Examples from literature

Please note, the following examples have been constructed from peer-reviewed publications and serve to contextualize specific aspects of microbiological risk assessment in an effort to guide the reader as to how the work published here can be applied in practice. The reader is reminded that also some of the examples make reference to specific national and value chain contexts; the data presented here are not suitable to generalize the findings beyond the context of the examples.

A1.1   SUMMARY OF THE EXAMPLES FROM LITERATURE

Examples	Key learnings	Reference chapter
1. Ice cream/ United States of America	• The outbreak illustrated that most consumers will not become ill when food contamination levels are low and growth does not occur, but emphasized the potential risk faced by highly susceptible persons when exposed to persistent contamination at potentially low levels. • The investigation highlighted the need to establish a stringent regime of environmental monitoring and product testing in RTE food production facilities, plus enhanced cleaning and sanitation, and heightened employee training. • The outbreak demonstrated that a blanket "zero tolerance" approach for all RTE foods provides a strong disincentive for product contact surface and end product testing. • The incident underlined the need for food processors to actively promote a culture of food safety and thoroughly identify and manage food safety hazards.	Exposure assessment; Hazard characterization; Monitoring

(cont.)

Examples	Key learnings	Reference chapter
2. Rockmelon/ Australia	• The outbreak highlighted the need to improve industry awareness of external threats to contamination of fresh produce – whole fruit grown near the ground in an open farm environment cannot be expected to be free from *L. monocytogenes*. • Essential to implement effective control measures such as washing and sanitizing of fruit and to validate the efficacy of sanitation procedures, monitor key variables such as sanitizer concentration, and introduce thorough environmental monitoring programmes in packing plants. • Additionally, there is a need for communication of potential risks to consumers and advice on hygienic handling and storage of fresh produce, especially to consumers in vulnerable groups.	Monitoring
3. Blood sausage/ Germany	• The identification of this outbreak and its vehicle resulted from an efficient collaboration between public health and food safety authorities in Germany and the European CDC to detect cross-border cases (one French case detected) (One Health Approach). • Use of conventional (cultural) methods was complemented by newer methods (MALDI TOF MS, specific clone outbreak PCR), and WGS helped to confirm and control the outbreak. • Strengthening surveillance in individual countries by harmonizing microbiological methods and providing epidemiologic tools for investigations will be a key step in reducing the public health burden of listeriosis, even as the population at risk grows.	Hazard characterization
4. Soft cheese/ Chile	• Lack of regulations contributed to this outbreak. Regulations were implemented and updated after the outbreak, with a significant improvement in the surveillance of *L. monocytogenes* for both foods and people. • Mandatory monitoring through submission of *L. monocytogenes* isolates to the national reference laboratory was crucial to notice the significant increase in the number of cases of listeriosis associated with this outbreak. The use of molecular techniques for the typing of isolates was key to findng the source of the outbreak. • Risk communication for populations at higher risk of acquiring listeriosis was significantly improved after the outbreak.	Monitoring

(cont.)

Examples	Key learnings	Reference chapter
5. Frozen vegetables/ Hungary	• The persistent presence of *L. monocytogenes* – this investigation revealed that *L. monocytogenes* persisted in the production environment for at least 3 years. • The extent of this outbreak might have been underestimated since it was identified through WGS, and only a subset of the European Union/ EEA countries routinely use it to characterize *L. monocytogenes* isolates. • Outbreak investigations showed that non-RTE foods can be used by the consumer as RTE foods; for example, frozen vegetables can be defrosted and used as such in salads by consumers, without undergoing any process to eliminate or reduce the level of pathogens.	Monitoring; Exposure assessment
6. Deli meat/ Canada	• The importance of trend analysis in tracking *Listeria* contamination in the processing environment. • The need of deep disassembly of equipment (mechanical slicer) prior to cleaning and sanitizing. • The value of educating health-care workers on the dangers associated with consumption of deli meat by the elderly and other high-risk populations. • The benefits of potential preventive controls such as post-package pasteurization and/or inclusion of *Listeria* growth inhibitors in the product formulation.	Monitoring
7. Polony/South Africa	• The outbreak illustrated how widespread contamination and inadequate surveillance of a food-processing environment can result in a massive outbreak of listeriosis. • How a lack of guidance on good hygienic practices and sampling programme at a frequency sensitive enough to detect contamination must be addressed. • A shift in attention from over-reliance on testing as proof of safety to environmental testing and use of post-processing treatments for product and external packaging has been adopted by some FBOs with success. • Beyond continuous improvements in food hygiene, authorities must provide industry with the regulatory science upon which to improve management and reduce cross-contamination risks. The quality of official food inspection is a critical factor.	Lab methods and sampling; Exposure assessment; Monitoring

A1.2   EXAMPLE 1: ICE CREAM

Ice cream outbreak | United States of America 2015
Outbreak

A complex multistate outbreak of listeriosis occurred in the United States of America, resulting in ten cases over a five-year period. Laboratory testing, epidemiological analysis, and traceback linked the outbreak to ice cream. The company issued two trade level recalls of selected products in March 2015, followed by a voluntary recall of all products from all company manufacturing facilities on 20 April 2015 after products produced in the Texas and Oklahoma facilities were identified as the source of the outbreak.

Confirmed cases: Ten people (with three deaths) all received implicated product while hospitalized.

Cases: Reported from four states: Arizona (1), Kansas (5), Oklahoma (1), and Texas (3)

Illness onset: Between January 2010 through January 2015

Investigation

Initially a cluster of five people hospitalized at the same Kansas hospital for unrelated problems developed invasive listeriosis. These patients were considered vulnerable, and each was infected with one of four strains of *L. monocytogenes*. Isolates from four of the patients were highly related by WGS. Illness onset dates for the five patients ranged from January 2014 through January 2015. Information available at the time indicated that certain ice cream products from Texas were the likely source of this outbreak. Hospital patients consumed multiple milkshakes made with ice cream as an ingredient.

Subsequent testing by State Departments of Health confirmed the presence of *L. monocytogenes* in unopened packages of ice cream leading to the series of recalls.

A retrospective review of the PulseNet database for PFGE matching isolates collected from ice cream samples resulted in the identification of further cases of illness in Arizona, Oklahoma and Texas with the onset period from 2010–2014.

Summary of findings

An unprecedented high number of ice cream products from the Texas facility were found to be contaminated with *L. monocytogenes*. The pathogen was detected in 99 percent (2 307 of 2 320) of samples (lower limit of detection, 0.03 MPN/g) manufactured between November 2014 and March 2015. Over 92 percent of samples were contaminated at < 20 MPN/g.

Analysis of the ten cases of listeriosis indicates the outbreak represents two separate events, as contaminated ice cream originating from different production facilities in Texas and Oklahoma.

While the ice cream products contaminated with *L. monocytogenes* were widely distributed to consumers, there were no identified cases among the public or among pregnant women. The uniform and low-level contamination of the ice cream products indicates that practically all consumers who consumed the products had exposure to the outbreak-associated strain. However, a detailed examination of the outbreak strongly suggests that all known exposures related to this outbreak were likely due to the consumption of milkshakes, rather than to the original ice cream product (Farber *et al.*, 2021).

Root cause analysis established that *L. monocytogenes* entered the production facilities through various sources and established harborages on equipment and in drains. As a result, facility-wide remediation efforts were implemented to prevent the reintroduction and re-establishment of *L. monocytogenes* in the facilities.

Key learnings

The investigation of the outbreak illustrated that most consumers will not become ill when food contamination levels are low and no growth is facilitated. But the outbreak further emphasized the potential risk faced by highly susceptible persons when exposed to persistent contamination at low levels, especially when a product not supporting growth is transformed into another product which has the potential to support growth.

This outbreak highlighted the need to establish a stringent regime of environmental monitoring and product testing in RTE food production facilities, implement enhanced cleaning and sanitation, and introduce heightened employee training. Test and hold procedures are also applicable for a product which has a long shelf-life.

The incident also underlined the need for food processors to actively promote a culture of food safety and thoroughly identify and manage food safety hazards, especially if their product is consumed by highly susceptible consumers.

References

CDC. 2015. Multistate outbreak of listeriosis linked to blue bell creameries products (Final Update). In: *Centers for Disease Control and Prevention*. Atlanta, Georgia, USA. Cited 22 April 2022, www.cdc.gov/listeria/outbreaks/ice-cream-03-15/

Chen, Y.I., Burall, L.S., Macarisin, D., Pouillot, R., Strain, E., De Jesus, A.J., Laasri, A., Wang, H., Ali, L., Tatavarthy, A., Zhang, G., Hu, L., Day, J., Kang, J., Sahu, S., Srinivasan, D., Klontz, K., Parish, M., Evans, P.S., Brown, E. W., Hammack, T.S., Zink, D. & Datta, A.R. 2016. Prevalence and level of *Listeria monocytogenes* in ice cream linked to a listeriosis outbreak in the United States. *Journal of Food Protection*, 79(11): 1828–1832. https://doi.org/10.4315/0362-028X.JFP-16-208

Farber, J.M., Zwietering, M., Wiedmann, M., Schaffner, D., Hedberg, C.W., Harrison, M.A., Hartnett, E., Chapman, B., Donnelly, C.W., Goodburn, K.E. & Gummalla, S. 2021. Alternative approaches to the risk management of *Listeria monocytogenes* in low risk foods. *Food Control*, 123: 107601. https://doi.org/10.1016/j.foodcont.2020.107601

Pouillot, R., Klontz, K.C., Chen, Y., Burall, L.S., Macarisin, D., Doyle, M., Bally, K.M., Strain, E., Datta, A. R., Hammack, T.S. & Van Doren, J.M. 2016. Infectious dose of *Listeria monocytogenes* in outbreak linked to ice cream, the United States of America, 2015. *Emerging Infectious Diseases*, 22(12): 2113–2119. https://doi.org/10.3201/eid2212.160165

A1.3 EXAMPLE 2: ROCK MELON

Rockmelon outbreak | Australia 2018
Outbreak

A total of 22 human cases of listeriosis linked by WGS to a common source occurred in Australia between January and April 2018. Laboratory testing, epidemiological analysis, and traceback linked the outbreak to rockmelons (cantaloupes) originating from a single farm in New South Wales (NSW). A trade level recall was initiated on 28 February 2018, and consumers were advised to discard any rockmelon they may have purchased.

Confirmed cases: 22 comprising nine males and 13 females (with seven deaths and one miscarriage)

Age: Principally elderly persons, mean age 70 years (range of 0 to 94 years)

Cases: Reported from four Australian states – NSW (6), Victoria (8), Queensland (7), and Tasmania (1)

Illness onset: Between 17 January to 10 April 2018

Investigation

The NSW Food Authority undertook extensive microbiological swabbing of the packing facility and later sampled melons from wholesale and retail markets. Melons obtained at retail, wholesale, and a single swab of melons at the packing shed tested positive for *L. monocytogenes*, and the genetic sequence of these isolates matched clinical cases (serotype 4b; ST240). An environmental swab taken at the farm also tested positive for the outbreak strain.

Based on epidemiological data, the prevalence and contamination of *L. monocytogenes* on rockmelon surfaces was estimated to be most likely very low (< 100 CFU/g). Importantly, there may have been temperature abuse of sliced rockmelon in the home.

No rockmelons or environmental swabs from other farms tested positive for *L. monocytogenes*.

Summary of findings

Investigations at the farm and the packing facility found the grower was following industry best practices with rockmelons being washed, sanitized and packed under hygienic conditions.

The original source of contamination is unknown. However, adverse weather (localized storm over the farm) and subsequent dust storms prior to harvest of implicated batches may have introduced contamination into the packing facility. Plus, the soil load may have compromised washing and sanitation processes, resulting in a low level of *L. monocytogenes* being present on the fruit. The possibility that contamination originated from within the packing facility was not verified.

Inspections by the NSW Food Authority found there was also an opportunity for the introduction of *Listeria* after washing via contact with surfaces or equipment that may have been contaminated with *L. monocytogenes*. Dust blown from fans used to dry the fruit after washing, and porous cushioning material on packing tables that was not able to be effectively cleaned, were also identified as potential sources of contamination.

Subsequent modifications on the site included changes to equipment, the packing line, cleaning and sanitizing procedures as well as documentation. The farm has now introduced elevated levels of sanitizer in both the washing/scrubbing and sanitizing steps in the packing facility.

Key learnings

Whole fruit grown near the ground in an open farm environment cannot be expected to be free from *L. monocytogenes*. This outbreak highlighted the need to improve knowledge and awareness of external threats to rockmelon safety and to implement improved control measures such as enhanced washing and sanitizing of fruit and improved hygiene within packing facilities. The possibility of extreme weather events impacting fruit contamination needs to be considered, and appropriate remedial action identified.

A key learning is that the efficacy of steps such as sanitation must be validated, and that key variables such as sanitizer concentration need to be monitored, along with implementing an effective environmental monitoring programme.

There also needs to be communication of the potential risks to consumers, and advice on hygienic handling and storage of rockmelons, especially after they have been cut. This is especially important for consumers in vulnerable groups, and Food Standards Australia New Zealand have added rockmelons to the list of foods to avoid because of the higher risk of contamination.

Reference

NSW Department of Primary Industries. 2018. *Listeria outbreak investigation – summary report for the melon industry*. New South Wales, Australia. ISBN: 978-1-76058-267-8

A1.4   EXAMPLE 3: BLOOD SAUSAGE

Blood sausage outbreak | Germany 2018–2019
Outbreak

An unusually large cluster of *L. monocytogenes* isolates was identified and named "Epsilon1a" outbreak, including 134 highly clonal, benzalkonium-resistant sequence type 6 (ST6) isolates collected from 112 notified listeriosis cases from 2018 to 2019. The outbreak was one of the largest reported in Europe in over 25 years. Cases occurred in 11 out of 16 federal states in Germany; most cases occurred in western and southern Germany. Epidemiologic investigations identified blood sausage contaminated with *L. monocytogenes* as being highly related to the clinical isolates. The outbreak ended after withdrawal of the product from the market.

Confirmed cases: 112 cases (with two deaths). Among case-patients, 90 percent reported consuming minced meat, and 80 percent reported consuming blood sausage.

Age: 111 case-patients were 53 to 98 (median 79) years of age; 66 (59 percent) were men; 45 (41 percent) were women. Seven (6.3 percent) case-patients died, two of whom had listeriosis as the primary cause of death. Only one (0.8 percent) pregnant woman was involved.

Cases: Cases occurred in 11 out of 16 federal states in Germany; most cases occurred in western and southern Germany.

Illness onset: Disease onset from August 2018 to June 2019

Investigation
- *L. monocytogenes* was isolated from 184 specimens from human cases and food sources.
- Isolation and enumeration from food samples was carried out according to ISO methods EN ISO11290–1:2017 and EN ISO 11290–2:2017.
- The species was identified by using EN ISO 11290–1:2017 or matrix-assisted laser desorption/ionization time-of-flight mass spectrometry and multiplex PCR.
- Isolates from listeriosis cases and suspected food samples were investigated using whole genome sequencing. Subtyping was carried out by core genome multi-locus sequence typing (cgMLST, Ruppitsch's scheme) and single nucleotide polymorphism.
- Virulome, resistome, and microdilution-based antimicrobial drug susceptibility was examined.

- Once the outbreak was identified, patients were interviewed by using a standardized questionnaire on their food consumption for two weeks before illness onset regarding general eating habits and food purchasing behaviors. These data identified 40 food items for inclusion in the case-control study.

Summary findings
- The case-control study with 41 case-patients and 155 controls reported purchase of food in a single specific supermarket chain by 98 percent of the case-patients, compared with 64.3 percent of the controls.
- A statistically strong association between cases and consumption of minced meat was detected. Among the case-patients, 90 percent reported consuming minced meat, and 80 percent reported consuming blood sausage, compared with 23 percent of the controls who consumed minced meat and 45 percent who consumed blood sausage.
- None of the case-patients were vegetarians.
- Purchases in a particular supermarket chain and blood sausage consumption were strongly associated with listeriosis in the case-control study.
- The outbreak clone was identified in blood sausage samples from a patient's household and from the linked supermarket chain.
- Investigation of the sliced blood sausage purchased at the implicated supermarket chain showed the highest contamination ($> 3 \times 10^6$ CFU/g). However, the amount of *L. monocytogenes* found in unopened blood sausage samples was below the limit of 100 CFU/g.
- A total of 134 highly clonal, benzalkonium-resistant, MLST sequence type 6 *L. monocytogenes* strains were isolated from 112 notified listeriosis cases. Only one (0.8 percent) pregnant woman was involved. This is likely due to the impact of official recommendations for this subpopulation in Germany. No gastroenteritis was reported.
- Of the 134 isolates, 99 were from blood samples, 13 from cerebrospinal fluid, and one each from lymph nodes, ascites, sputum, pleura, joints, abscesses or a superficial wound.
- The virulome analysis revealed Listeria pathogenicity island 1 (LIPI-1) in all outbreak isolates and LIPI-3 in 64 percent of them. A phylogenetically diverse cluster was identified by using cgMLST. "Epsilon1" included 46 PCR serogroup Ivb isolates belonging to ST6, which could be further subtyped into 12 cgMLST complex types.
- Resistome analysis demonstrated the prevalence of the emrC gene, which is associated with benzalkonium chloride tolerance. Susceptibility testing revealed sensitivity to most clinically relevant antimicrobial drugs, but all tested isolates were fully resistant to ceftriaxone and daptomycin.

Key learnings

- The identification of this outbreak and its vehicle resulted from an efficient collaboration between public health and food safety authorities in Germany and the European CDC to detect cross-border cases (one French case detected).
- Geographically widespread outbreak due to long distance food-trade connections and travel.
- Systematic collection of food isolates, a continuous exchange of information, and a WGS-based subtyping methodology by food safety authorities played an important role.
- Due to the long incubation time, severity of the illness or death, it was difficult to have the interviews and questionnaire with the patients.
- Remarkably homogeneous cluster with 0–5 (median 0) different cgMLST alleles (threshold ≤ 10 alleles) suggesting that the clone only persisted in the production facility and likely did not multiply.
- Storage beyond the anticipated shelf-life or insufficient refrigeration likely resulted in suspected growth of *L. monocytogenes* on the blood sausages.
- The outbreak demonstrates how WGS-based pathogen surveillance combined with efficient interventions of the involved stakeholders can improve the management and prevention of foodborne diseases such as listeriosis.
- Use of conventional (cultural) methods complemented by emerging (MALDI TOF MS, specific clone outbreak PCR) technologies and WGS helped to confirm and control the outbreak.
- Strengthening surveillance in individual countries by harmonizing microbiological methods and providing epidemiologic tools for investigations is a key step in reducing the public health burden of listeriosis, even as the population at-risk grows (Hedberg, 2006).

References

Halbedel, S., Wilking, H., Holzer, A., Kleta, S., Fischer, M.A., Lüth, S., Pietzka, A., Huhulescu, S., Lachmann, R., Krings, A., Ruppitsch, W., Leclercq, A., Kamphausen, R., Meincke, M., Wagner-Wiening, C., Contzen, M., Kraemer, I.B., Al Dahouk, S., Allerberger, F., Stark, K. & Flieger, A. 2020. Large nationwide outbreak of invasive listeriosis associated with blood sausage, Germany, 2018-2019. *Emerging Infectious Diseases*, 26(7): 1456–1464. https://doi.org/10.3201/eid2607.200225

Hedberg, C. 2006. *Listeria* in Europe: the need for a European surveillance network is growing. *Eurosurveillance*, 11(6): 75–76.

A1.5   EXAMPLE 4: SOFT CHEESES

Cheese outbreak | Chile 2008
Outbreak

A listeriosis outbreak involving 78 cases occurred in Chile in 2008. This outbreak was linked to the consumption of soft cheeses, specifically Brie and Camembert, which were manufactured by the same milk processor, but sold under different brand names. Cheeses were recalled from the market on 25 November 2008, with a subsequent reduction in the number of cases in the following two weeks. No further listeriosis cases caused by the epidemic strain, *L. monocytogenes* clone 09 (serogroup 4b, ST1, and CC1), were reported during the last two weeks of December 2008, after recall of the products. Only three sporadic cases caused by this strain were reported in 2009.

Confirmed cases: a total of 78 cases, affecting mostly pregnant women. Neither the number of males or females, nor the age of the cases were described in the official reports of the outbreak investigation at that point.

Geographical distribution of cases: This outbreak affected 8 out of 13 regions of Chile. Most cases (65; 83.3 percent) were reported in the metropolitan region of Santiago, 7 (9 percent) in Valparaiso region, and the remaining 6 (7.7 percent) cases distributed among six regions; namely Araucanía, Maule, Bío-Bío, Atacama, O´Higgins, and Los Ríos regions. Most of the cases from the metropolitan region were residents from high-income urban areas.

Illness onset: January to December 2008.

Fatalities: 14 deaths, corresponding to elderly people with underlying medical conditions, immunocompromised individuals, and a newborn. *Specific numbers for each category were not provided by the official reports.

Investigation

During 2008, a significant increase in the number of listeriosis cases was observed in Chile (a five-fold increase as compared with previous years). In September 2008, the health authorities of the metropolitan region started an epidemiological investigation of food products possibly associated with the outbreak, using food consumption data collected from the clinical cases. During the investigation, *L. monocytogenes* Clone 09 was isolated from soft-cheese samples of the same name brand which had been obtained from the refrigerators of two different cases. This finding led the investigation towards the cheese processor premises, where further soft-cheese samples were collected and tested positive for the same *L. monocytogenes* Clone 09 strain. Eventually, the soft cheeses sold under three different name brands but manufactured by the same processor were recalled from the markets on 25 November 2008. No further details regarding the epidemiological investigation and findings at the processor were disclosed.

Key Learnings

At the time of this outbreak, Chile was only under laboratory surveillance for *L. monocytogenes* (that is mandatory submission of isolates to the National Reference Laboratory), without mandatory notification of human clinical cases. Laboratory surveillance has been complemented, starting in 2020, with mandatory notification of all human listeriosis cases detected in Chile. This improvement was a result of changes and updates in the law that regulates communicable diseases.

Before the described 2008 listeriosis outbreak, microbiological criteria for *L. monocytogenes* in foods was not included in the Chilean Food Code (*Reglamento Sanitario de los Alimentos*). This situation changed in 2009, triggered by the 2008 listeriosis outbreak.

Changes and implementation of new regulations were established for both notification of listeriosis and microbiological criteria for *L. monocytogenes* in foods. Furthermore, these changes allowed for an improvement in surveillance and data collection of listeriosis cases, as well as in the risk communication of listeriosis in Chile. All these improvements, along with the routine use of molecular techniques for confirmation and typing of *L. monocytogenes* in the National Reference Laboratory (ISP), have allowed for faster response in case investigations and early detection of potential outbreaks.

References

Ministerio de Salud de Chile, Departamento de Epidemiología. 2009. *Informe Brote Listeriosis Región Metropolitana.*

Ministerio de Salud, Chile, Departamento de Epidemiología. 2010. *Informe de Listeria, Chile 2010.*

Toledo, V., den Bakker, H. C., Hormazábal, J. C., González-Rocha, G., Bello-Toledo, H., Toro, M. & Moreno-Switt, A. I. 2018. Genomic diversity of *Listeria monocytogenes* isolated from clinical and non-clinical samples in Chile. *Genes,* 9(8): 396. https://doi.org/10.3390/genes9080396

Ulloa, S., Arata, L., Alarcón, P., Araya, P., Hormazabal, J.C. & Fernandez, J. 2019. Caracterización genética de cepas de *Listeria monocytogenes* aisladas durante los años 2007-2014 en Chile. *Revista Chilena de Infectología,* 36(5): 585–590.

Frozen vegetables outbreak | Europe 2015-2018

Outbreak

A multicountry outbreak of *L. monocytogenes* ST6 that caused 54 cases and ten deaths over the period from 2015 to 2018 was linked in 2018 to frozen vegetables. The beginning of the investigation into the frozen vegetables began in France, where researchers isolated a *L. monocytogenes* strain from a conveyor belt that moves frozen vegetables along; the strain which was isolated matched the outbreak strain. In total, WGS analysis of 29 non-human *L. monocytogenes* isolates found them to be closely related to the multicountry human cluster of *L. monocytogenes* PCR serogroup IVb, MLST sequence type 6 (ST6). The WGS analysis provided a strong microbiological link between the human and the nonhuman isolates, and this was indicative of a common source related to frozen corn and other frozen vegetable mixes persisting in the food chain. Environmental contamination of a vegetable freezing plant was indicated as the source of the persistence of the strain causing the outbreak from 2015 until 2018. Five MS were involved but implicated frozen products, some of them with a long shelf-life (some until mid-2020), which were distributed to 116 countries.

Confirmed cases: 53 (females and males); case-fatality rate around 20 percent

Cases reported in six countries: Austria, Australia, Denmark, Finland, Sweden and the United Kingdom

---

Investigation

Traceability information for the frozen corn samples pointed to frozen corn products packed in Poland and processed and produced in a freezing plant in Hungary. Since *L. monocytogenes* IVb ST6 matching the outbreak strain was isolated from frozen spinach and frozen green beans sampled at the Hungarian plant, it is possible that frozen vegetables, other than corn, which was processed in this plant, could also be implicated as a vehicle of human infection. The information available confirmed contamination within the Hungarian processing plant.

During the summer of 2018, the company, in full collaboration with the authorities and independent experts, conducted a large in-depth review of the facility to identify the root cause of the potential contamination. Investigation revealed that a persistent presence of *L. monocytogenes* was found in one of the two freezing tunnels and as a result, the tunnel at the Baja-based plant was closed. The rest of the freezing plant was cleared after inspection in cooperation with the food safety authorities and in September 2018, production was restarted.

Outbreak investigations conducted showed that some frozen vegetables can be defrosted and used as such in salads by consumers or as ingredients in other ready-to-eat (RTE) products subsequently sold to consumers without undergoing any process to eliminate or reduce the level of pathogens. If such defrosted fruits, vegetables or herbs (FVH) are stored for a prolonged period at refrigeration temperatures, the potential growth of *L. monocytogenes* could represent a serious public health risk.

Summary of findings

A multicountry outbreak of *L. monocytogenes* ST6 that caused 54 cases and ten deaths over the period from 2015 to 2018 was linked in 2018 to frozen vegetables. Environmental contamination of a freezing plant was indicated as the source of the persistence of the strain causing the outbreak. Five MS were involved, but implicated frozen products were distributed to 116 countries and some of these products had a long shelf-life (some until mid-2020). Investigations revealed that a persistent presence of *L. monocytogenes* was found in one of the two freezing tunnels and as a result, the tunnel at the Baja-based plant was closed.

Key learnings

The WGS analysis provides a strong microbiological link between the human and non-human isolates, and this is indicative of a common source related to frozen corn and other frozen vegetable mixes including corn persisting in the food chain. Data from WGS provided evidence to detect the outbreak.

New consumer preferences might have played a role in this outbreak, as investigations conducted showed that some frozen FVH can be defrosted and used as such in salads by consumers or as ingredients in other RTE products subsequently sold to consumers, without undergoing any process to eliminate or reduce the level of pathogens.

This outbreak highlighted the need to establish a stringent regime of environmental monitoring and product testing, test-and-hold procedures, enhanced cleaning and sanitation, and heightened employee training in fresh and frozen vegetable processing plants. This is not fully implemented in all countries. Furthermore, in general, frozen vegetables should i) be considered as non-RTE foods, ii) be labelled with adequate cooking instructions, and iii) be cooked prior to consumption. Food safety education and guidance regarding such cooking instructions on frozen vegetables, especially for at-risk consumers, should be a priority.

References

EFSA. 2020. The public health risk posed by *Listeria monocytogenes* in frozen fruit and vegetables including herbs, blanched during processing. *EFSA Journal*, 18(4): 6092. https://efsa.onlinelibrary.wiley.com/doi/epdf/10.2903/j.efsa.2020.6092

EFSA & ECDC. 2018. Multi-country outbreak of *Listeria monocytogenes* serogroup IVb, multi-locus sequence type 6, infections linked to frozen corn and possibly to other frozen vegetables – first update. *EFSA Supporting Publication*: EN-1448. https://efsa.onlinelibrary.wiley.com/doi/epdf/10.2903/sp.efsa.2018.EN-1448

Deli meat outbreak | Canada 2008

Outbreak

Fifty-seven individuals across seven Canadian provinces (Ontario – 41, Quebec – 5, British Columbia – 5, Alberta and Saskatchewan – 2 each, Manitoba and New Brunswick – 1 each) developed listeriosis from 3 June to 22 November 2008. Epidemiological analysis, traceback, and PFGE analysis linked the outbreak to delicatessen meats.

Confirmed cases: 57 (24 deaths – average age 76 years)

Age range: 29 to 98 years (average 74 years)

Gender: 33% male, 67% female (no maternal/neonatal cases)

Cases: At least 41 had underlying medical or immunocompromising conditions; 86% of cases resided in long-term care facilities or were hospital inpatients/outpatients.

Illness onset: 3 June to 22 November 2008

A total of 191 meat products were recalled nationwide.

Investigation

Investigators documented a total of 57 cases of listeriosis from a series of small clusters from multiple long-term and acute care facilities and additional surveillance data. Food consumption data from the case followed by traceback investigations implicated deli meat as the source of the outbreak. Production Lines A and B that produced pre-sliced deli meats for hospitals, long-term care facilities, prisons, hotels and restaurants yielded food contact surfaces that were positive for *Listeria* spp. both before and during the outbreak. *L. monocytogenes* was subsequently recovered from 82 of 163 deli meat samples of six different types (levels of < 50 to > 20 000 CFU/g) produced on Lines A and B from 12 June to 2 August. All *L. monocytogenes* isolates, which belonged to serotype 1/2a and three closely related PFGE patterns were indistinguishable from the clinical isolates, thereby confirming the source of the outbreak. Subsequent work demonstrated the presence of a novel *Listeria* genomic island (LGI1) which afforded the outbreak strain some resistance to the sanitizer benzalkonium chloride – a quaternary ammonium compound. It is unknown, however, if this resistance played any role in this outbreak.

Summary of findings

After several nationwide recalls were issued in mid-August, production eventually resumed in September after deep cleaning and sanitizing. However, cleaning and sanitizing practices were inadequate, with employee movement also creating opportunities for finished product contamination. The outbreak strains were still recovered from Lines A and B during September and October, suggesting long-term colonization and persistence in production areas.

Key learnings

Factors contributing to the establishment of *L. monocytogenes* in the facility likely included plant construction and condensation issues, cross-contamination of finished product between lines, and inadequate deep cleaning and sanitation (biofilm formation), with cross-contamination from mechanical slicers being the most likely cause. The company did not conduct the trend analysis required under its *Listeria* control policy. In addition, employees in the plant were not required to, nor did they volunteer, any information concerning the repeated findings of *L. monocytogenes* in the plant to CFIA inspectors. Finally, company staff had notified their superiors of the repeated presence of *L. monocytogenes*, but this information was not sent to the Head Office because it was erroneously thought that the plant's interventions had controlled the problem. This outbreak also highlights the lapses in rapidly identifying cases of listeriosis and in educating health-care workers about the dangers associated with the consumption of deli meat by the elderly and other high-risk populations.

References

Currie, A., Farber, J.M., Nadon, C., Sharma, D., Whitfield, Y., Gaulin, C., Galanis, E., Bekal, S., Flint, J., Tschetter, L., Pagotto, F., Lee, B., Jamieson, F., Badiani, T., MacDonald, D., Ellis, A., May-Hadford, J., McCormick, R., Savelli, C., Middleton, D., Allen, V., Tremblay, F.-W., MacDougall, L., Hoang, L., Shyng, S., Everett, D., Chui, L., Louie, M., Bangura, H., Levett, P.N., Wilkinson, K., Wylie, J., Reid, J., Major, B., Engel, D., Douey, D., Huszczynski, G., Lecci, J.D., Strazds, J., Rousseau, J., Ma, K., Isaac, L. & Sierpinska, U. 2015. Multi-province listeriosis outbreak linked to contaminated deli meat consumed primarily in institutional settings, Canada, 2008. *Foodborne Pathogens and Disease*, 12(8): 645–652. https://doi.org/10.1089/fpd.2015.1939

Kovacevic, J., Ziegler, J., Walecka-Zacharska, E., Reimer, A., Kitts, D.D. & Gilmour, M.W. 2016. Tolerance of *Listeria monocytogenes* to quaternary ammonium sanitizers in mediated by a novel efflux pump encoded by emrE. *Applied and Environmental Microbiology*, 82: 939–953.

Government of Canada. 2000. Report of the independent investigator into the 2008 listeriosis outbreak. In: *Government of Canada*. Ottawa, Canada. Cited 5 July 2021. www.canada.ca/en/news/archive/2009/07/report-independent-investigator-into-2008-listeriosis-outbreak.html

Polony outbreak | South Africa 2017-2018
Outbreak

An outbreak of listeriosis commenced across South Africa in 2017. Investigations in early 2018 traced the outbreak strain to an RTE processed meat product, polony. Most cases were reported from Gauteng Province (58 percent) followed by Western Cape (13 percent) and KwaZulu-Natal (eight percent) provinces. Cases were diagnosed in both public (64 percent) and private (36 percent) healthcare sectors.

Confirmed cases: 1060 (216 deaths)

Cases associated with pregnancy (465)

Pregnancy-associated cases that occurred in neonates (406)

Age range: 15 to 49 years (for 937 cases excluding those who were pregnant)

Age range from birth to 93 years (median 19 years)

Neonates aged ≤ 28 days accounted for 43 percent of the cases, and 95 percent had early-onset disease (birth to ≤ 6 days)

Patients with known human immunodeficiency virus (HIV) status: (38 percent of the pregnancy-associated cases; 46 percent of the general patients)

Gender: 55 percent female; illness onset: unknown

A total of 5 000 tons of product was recalled from 15 countries.

No outbreak was reported, and *L. monocytogenes* strains were not identified in the 14 countries which imported RTE processed meat products.

Other important factors

The world's top ten countries with the highest HIV rates are in Southern Africa, and South Africa (RSA) is at number one.

RSA Health Care Index rank in 2020 (48) and UHC effective coverage index 2019 (59.7)

Top 3 deaths causes of mortality in RSA; HIV (1); lower respiratory infections (2); tuberculosis (3)

Percentage of unwanted births in 2016: (20.4 percent)

South Africa literacy rate for 2017 was 87.05 percent and South Africa has 11 official languages (product labelling?)

Investigation

In July and August 2017, clinicians and microbiologists at a number of sites in one province in South Africa reported an increase in cases of neonatal sepsis and adult meningitis due to *L. monocytogenes*. A review of laboratory-confirmed cases in the public and private sector confirmed that there was a dramatic increase in the weekly number of cases. Initial investigations included establishment of surveillance networks, collection of clinical isolates of *L. monocytogenes* for WGS, development and distribution of case investigation forms, consultative development of diagnostic and treatment guidelines, engagement with the food sector to obtain isolates from food, and environmental quality control specimens for WGS. Outbreak investigations conducted included 1) detailed food history interviews amongst cases with laboratory-confirmed listeriosis; 2) microbiological culture of food obtained from the fridges of persons with laboratory-confirmed listeriosis; 3) surveillance for clusters of febrile gastro-enteritis; and 4) WSG of clinical, food and environmental isolates of *L. monocytogenes*. The outbreak strain, a strain of sequence-type 6 (ST-6) was found to be present in over 92 percent of cases where isolates were available for testing.

By early January 2018, food history interviews suggested that polony was amongst the most commonly consumed foodstuff amongst persons with listeriosis. A cluster of cases of febrile gastro-enteritis was reported on 13 January 2018, when nine children presented to a hospital. The diagnosis of listeriosis was confirmed through culture of the bacterium in a stool specimen from one of the children. Specimens from two different brands of polony obtained from the crèche attended by the children yielded growth of an ST6 *L. monocytogenes*. Environmental swabbing and food testing from specimens obtained from two manufacturers yielded *L. monocytogenes*. The ST6 was found in the environment and in polony samples after an extensive inspection of one manufacturer's production facility, and WGS results and analysis confirmed the presence of the outbreak strain on 3 March 2018.

Summary of findings

After closure of the establishment and recalls and destruction of 5 000 tons of product, production eventually resumed after a redesign of hygienic zones, deep cleaning and sanitizing. In affected establishments, the introduction of raw meat was identified as the main potential source of contamination of the establishment, followed by widely disseminated contamination in the entire food-processing environment. Cleaning and sanitizing practices were inadequate with a possibility of employee and equipment movement also creating opportunities for finished product contamination. The outbreak strain could not be recovered from the post heat treatment areas and RTE products after corrective actions and reopening.

During the outbreak, the levels of *L. monocytogenes* reported from the respective RTE product was below 10 CFU/g.

Key learnings

Deficiencies associated with listeriosis cases and the outbreak included a lack of adequate GMPs, poor maintenance of hygienic zones in the facility, inadequate monitoring, absence of a food safety culture and limited or ineffective regulatory oversight. Models of national and local food safety control systems are vital as multiple competent authorities and different standards on the same products bring confusion and gaps, compliance uncertainty and legislative fragmentation. Lack of a sensitive surveillance programme to detect low prevalence of *L. monocytogenes* at control points and the absence of a risk-based environmental monitoring programme coupled with a representative testing frequency sensitive enough to detect contamination has negative effects. The capability to characterize *L. monocytogenes* is important, as it assists with source attribution and generation of more data about the organism. A whole chain approach is important in the control of *L. monocytogenes*, as some strains are introduced into RTE establishments through contaminated raw materials. Reliance on private standards certification systems is not a guarantee for quality assurance and may give false assurance to both competent authorities and FBOs. Quality and effectiveness of cleaning chemicals is also important. Food is not manufactured in a sterile environment, hence understanding risk factors and adherence to GHP, good food safety culture practice, effective training of competent inspectors and quality managers, proper general cleaning and disinfection, effective periodic deep cleaning and ability to detect and eliminate any contamination, are all important interventions. Guidelines for an acceptable facility design from a sanitary point is a critical factor.

Reference

Smith, A.M., Tau, N.P., Smouse, S.L., Allam, M., Ismail, A., Ramalwa, N.R., Disenyeng, B., Ngomane, M. & Thomas, J. 2019. Outbreak of *Listeria monocytogenes* in South Africa, 2017-2018: laboratory activities and experiences associated with whole-genome sequencing analysis of isolates. *Foodborne Pathogens and Disease*, 16(7): 524–530. https://doi.org/10.1089/fpd.2018.2586

An analysis of data from listeriosis outbreak investigations

A2.1 METHODS

Internet searches were conducted using the search terms "*Listeria* outbreak" and "listeriosis outbreak" from March to October 2020. Similar searches were conducted using research engines, such as Web of Science, PubMed, ProMED, and Google Scholar. Competent authority websites were also scanned for information on outbreaks related to *Listeria*. Media sites, such as Food Safety News were also scanned and monitored for reports of past and on-going outbreaks of *Listeria*.

Only outbreaks of invasive listeriosis that identified a strong connection to a food source were captured as part of this review. This strong connection could either be through an epidemiological or food safety investigation of a combination of both. All data, including the year, the countries involved, the food and contamination source, the serotype, enumeration results and information on the clinical cases was captured using an Excel spreadsheet if available.

Many outbreaks spanned the course of several years or were confirmed retrospectively using molecular methods. In these situations, outbreaks were documented based on the year in which the outbreak source was determined. Outbreaks involving multiple countries were counted as one outbreak event; however, if the outbreak spanned more than one WHO region, it was double counted as an outbreak for each WHO region. If some or all outbreak cases were linked to hospital exposure (nosocomial) or exposure from a long-term care or convalescent care centre, they were each recorded as a "hospital (H) outbreak".

The food source was divided into the following categories: RTE Meat, RTE Dairy, RTE Fish and Seafood, RTE Fruits and Vegetables, RTE Egg and RTE Multiple Foods. The Multiple Food category was used for complex foods made up of many ingredients, such as sandwiches, if the outbreak investigation did not identify an ingredient of the food responsible for the contamination or the original source of contamination in situations involving cross-contamination. Any additional food details deemed relevant to the outbreak investigation, such as product variety or process descriptors, were captured as Food Details.

If any of the information was not available as part of the literature search, the field was left blank.

A2.2 RESULTS

L. monocytogenes outbreaks attributed to a food source were identified for 23 countries, spanning four WHO geographic regions: AMR, WPR, EUR and AFR, for the period between 2005 and 2020 (Table A1 and Figure A1). In total, 127 reported outbreaks of listeriosis linked to a particular food source were found; 69 (54 percent) occurred within the European region (EUR), 49 (38 percent) within the Americas region (AMR), 9 (7 percent) in the Western Pacific region (WPR) and one (0.8 percent) in the African region (AFR). One outbreak linked to mushrooms in 2019 spanned two WHO regions, AMR and WPR.

Competent authorities from China and Japan confirmed that there have been no outbreaks of foodborne listeriosis in the period of interest. Japan reported one suspected outbreak of foodborne listeriosis linked to cheese in 2001. Country specific searches were conducted, but no reported outbreaks of foodborne listeriosis were identified in either the Eastern Mediterranean (EMR) or the South-East Asia (SEAR) regions.

The outbreaks recorded a total of 3 628 cases of invasive listeriosis, of which at least 606 (17 percent) of the cases were reported as maternofoetal, and 230 (6 percent) as immunocompromised. In total, 554 (15 percent) of the cases resulted in death, of which at least 27 (5 percent) were perinatal. Foodborne exposure within a hospital or long-term/convalescent care environment was reported in 22 (17 percent) of outbreaks.

Of these 127 outbreaks, 40 (31 percent) were linked to RTE meat products, 36 (28 percent) to RTE dairy products, 17 (13 percent) to RTE fresh or minimally processed fruit and vegetable products, 15 (12 percent) to RTE fish and seafood products, and there was one (0.8 percent) linked to hard-boiled eggs. Multiple foods or foods with multiple ingredients (i.e. sandwiches, hummus, rice pudding, etc.) contributed to 18 (14 percent) of the total number of listeriosis outbreaks.

In the AMR, most outbreaks of listeriosis were linked to RTE dairy products, whereas in the EUR, RTE meat products were the most likely source. In the WPR, both RTE meat and fresh produce were both equally the most likely foods to be linked to a listeriosis outbreak. The AFR reported one very large outbreak linked to RTE meat.

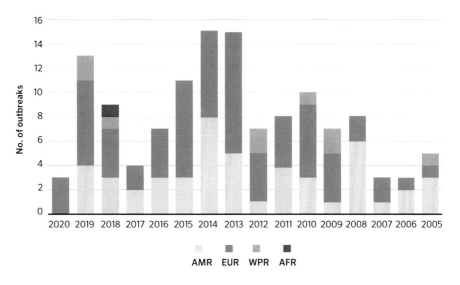

FIGURE A1. Outbreaks of listeriosis attributed to a food source in WHO regions from 2005–2020

Source: Author's own elaboration.

In the WPR and AFR, all reported outbreaks were associated with at least one reported death, while in the AMR only 65 percent and 55 percent in the EUR were linked to at least one death. RTE meat and dairy products were implicated in 64 percent of the listeriosis outbreaks associated with reported deaths.

RTE meat and dairy products are most often associated with an outbreak of listeriosis with reported foodborne exposure in a hospital setting. In the AMR, only 12 percent of outbreaks reported hospital foodborne exposure, compared to 20 percent in the EUR and 22 percent in the WPR.

TABLE A1. Number and proportions of outbreaks of listeriosis attributed to RTE food commodities in WHO regions

	AMR		EUR		WPR		AFR		
	Number	%	Number	%	Number	%	Number	%	Total
RTE meat	9	18	27	39	3	33	1	100	40
RTE dairy	21	43	13	19	2	22	-	-	**36**
RTE fish & seafood	-	-	14	20	1	11	-	-	**15**
RTE fruit & vegetables	12	24	3	4	3	33	-		**18 (17)***
Multiple foods	6	12	12	17	-	-	-	-	**18**
RTE egg products	1	2	-	-	-	-	-	-	**1**
Total	**49**		**69**		**9**		**1**		**128 (127)***
Outbreaks associated with deaths									
RTE meat	5	16	17	45	3	33	1	100	**26**
RTE dairy	16	50	7	18	2	22	-	-	**25**
RTE fish & seafood	-	-	9	24	1	11	-	-	**10**
RTE fruit & vegetables	6	19	3	8	3	33	-	-	**12 (11)***
Multiple foods	4	13	2	5	-	-	-	-	**6**
RTE egg products	1	3	-	-	-	-	-	-	**1**
Total	**32**		**38**		**9**		**1**		**80 (79)***
Outbreaks associated with hospital exposure									
RTE meat	2	33	3	21	2	100	-	-	**7**
RTE dairy	2	33	4	29	-	-	-	-	**6**
RTE fish & seafood	-	-	-	-	-	-	-	-	**0**
RTE fruit & vegetables	1	17	1	7	-	-	-	-	**2**
Multiple foods	1	17	6	43	-	-	-	-	**7**
Egg products	-	-	-	-	-	-	-	-	**0**
Total	**6**		**14**		**2**		**0**		**22**

* One outbreak linked to mushrooms in 2019 spanned two WHO regions, AMR and WPR.
Source: author's own elaboration

A2.3 IMPLICATED FOOD COMMODITIES

Of the reported 40 outbreaks linked to RTE meat products, 48 percent of the products specifically referenced an RTE deli-type product, of which 63 percent were reported as being sliced. Of the remaining RTE meat products, 24 percent were reported as either being sliced or cut.

For RTE dairy, of the 36 reported outbreaks, 30 (83 percent) were associated with cheese, of which three (3) were confirmed unpasteurized, twelve (12) pasteurized and five (5) did not specify whether the cheese product was pasteurized or unpasteurized. Three outbreaks were linked to fluid milk, two (2) pasteurized and one (1) to unpasteurized, raw milk. Of the three reported outbreaks linked to milk, all three implicated flavoured milk, while one also mentioned white milk. In addition, ice cream was implicated in two hospital outbreaks, and sour cream in one outbreak.

Of the 15 outbreaks implicating RTE fish and seafood, 9 (60 percent) of the products specifically mentioned that the fish was smoked or gravid (cured using salt, sugar and dill).

Of the 17 RTE fruit and vegetable associated outbreaks, 11 (65 percent) implicated vegetables; leafy greens (3), sprouts (3), frozen vegetables (2), vegetables and juices and other products thereof – mixed salad (1), mushrooms (1) and celery (1). The six outbreaks implicating fruit included cantaloupes (3), caramel apples (2) and fresh stone fruit (peaches, nectarines, etc.).

Of the 18 outbreaks involving multiple foods, sandwiches (3), hummus (3), and prepared salads (4) were most implicated.

A2.4 CONTAMINATION SOURCE

When the source of *L. monocytogenes* contamination that led to the outbreak was reported, in most outbreaks (59 percent), the outbreak investigation determined that the food manufacturing facility (food-processing environment) was the cause. Many publications reported finding a similar strain of *L. monocytogenes* within the environment of the processing facility, on equipment or food contact surfaces or in samples of unopened product. For processed RTE products, the contamination occurred post-processing (i.e. conveyors belts, pumps, slicing and chopping equipment), prior to or during packaging. There were five (5) outbreaks that cited underprocessed contaminated ingredients as the source of *L. monocytogenes*, for example, surface contamination of whole cantaloupes from the field. There were also six (6) incidents where cross-contamination occurred at foodservice establishments. For 32 percent of the outbreaks, the source of the contamination was either not reported or undetermined within the literature or reference.

Within the publication or reference, the *L. monocytogenes* serotype was referenced for 70 (55 percent) of the outbreaks, of which 47 percent cited serotype 1/2a and 36 percent serotype 4b. Enumeration results associated with food product testing were also mentioned for 33 (26 percent) of the outbreaks. Of these, 64 percent of the food samples had enumeration of *L. monocytogenes* exceeding 100 CFU/g.

TABLE A2. Outbreaks of listeriosis with a confirmed linked to a food source reported from 2005–2020

Year *Multiyear outbreak	Country	Food category	Subcategory	Contamination source	Serotype	Enumeration results (food products)	Age range (median)	Total no. cases (invasive) H = Hospital outbreak	Cases Maternofoetal	Immuno -comp	No. of deaths[a]	No. perinatal deaths	Ref.
2020	France	RTE Dairy	Cheese – various	-	-	-	-	10	-	-	-	-	FSN, 2020a; Chailloux, 2020
2020	Netherlands	RTE Fish and Seafood	Trout – fillets (chilled smoked)	-	-	-	42-85 (78)	6	-	-	2	-	Whitworth, 2020a
2020	Switzerland	RTE Dairy	Cheese – semi-hard (pasteurized)	Processing contamination	-	-	66-86	34H	-	-	10	-	FSN, 2020b
2019	Australia	RTE Fish and Seafood	Salmon (smoked)	Processing contamination	-	-	>70	3	-	-	2	-	BBC News, 2019; ADOHA, 2019
2019*	Belgium, Netherlands	RTE Meat	Deli meats (sliced)	Processing contamination	4b	<100-270 000 cfu/g	30-94	35	3	-	6	2	EFSA and EDCD, 2019; FSN, 2020c; Whitworth, 2019b
2019	Canada	RTE Meat	Chicken (cooked, diced)	Processing contamination	-	-	51-97	7H	-	-	0	-	PHAC, 2019
2019*	Denmark	Multiple Foods	Hummus and salads	Processing contamination	-	-	30-91	11	-	-	0	-	FSN, 2020d
2019*	Denmark, Estonia, Finland, France, Sweden	RTE Fish and Seafood	Trout and salmon (cold smoked)	Processing contamination	-	<10-570 cfu/g	64-83 (76)	22	-	-	5	-	ECDC and EFSA, 2019

[a] Does not include foetal losses.

(cont.)

Year *Multiyear outbreak	Country	Food category	Subcategory	Contamination source	Serotype	Enumeration results (food products)	Age range (median)	Total no. cases (invasive) H = Hospital outbreak	Cases Maternofoetal	Immuno -comp	No. of deaths[a]	No. perinatal deaths	Ref.
2019*	Germany	RTE Meat	Deli meats (sliced)	Processing contamination	1/2[a]	-	31-91	37	0	-	3	0	DW News, 2019; Whitworth, 2019a
2019*	Germany	RTE Meat	Sausage - blood (sliced or packed)	Processing contamination (slicing or packing)	-	<100 cfu/g	53-98 (79)	112	1	-	7	-	Halbedel et al., 2020
2019	Spain	RTE Meat	Chilled roasted pork	Processing contamination	4b	>15 000 cfu/g	-	222	38	-	3	6	WHO, 2019
2019	United Kingdom	RTE Meat	Deli meat	Processing contamination	-	-	-	9H	-	-	6	-	PHE, 2019; FSA, 2019
2019*	United States of America	RTE Egg	Whole - hard boiled	Processing contamination	-	-	1-82 (75)	7	1	-	1	0	USFDA, 2019
2019	United States of America	Multiple Foods	Deli sliced cheese or meat	Processing contamination - cross contamination during slicing	-	-	44-88 (64)	10	-	-	1	-	CDC, 2019a
2019*	United States of America, Australia	RTE Fruit and Veg.	Fresh mushrooms - whole	-	-	-	-	42	-	-	4	-	Beach, 2020; USFDA, 2020; Whitworth, 2020b
2018	Australia	RTE Fruit and Veg.	Fresh cantaloupes - whole	Processing - ineffective washing	-	-	-	19	-	-	7	1	NSW DPI, 2018; WHO, 2018
2018	Austria	RTE Meat	Liver pâté	Post processing (food service)	4b	<10 cfu/g	4-84	3[b]	-	-	0	-	Cabal et al., 2019

[b] Ten cases of febrile gastroenteritis were also associated with this outbreak.

(cont.)

Year *Multiyear outbreak	Country	Food category	Subcategory	Contamination source	Serotype	Enumeration results (food products)	Age range (median)	Total no. cases (invasive) H = Hospital outbreak	Cases Maternofoetal	Immuno -comp	No. of deaths[a]	No. perinatal deaths	Ref.
2018	Canada	RTE Meat	Deli meats (sliced)	Post- processing contamination (food service)	-	-	-	7	-	-	0	-	SMDHU, 2018
2018*	Denmark, Germany, France	RTE Fish and Seafood	Salmon (smoked)	Processing contamination	-	110–240 cfu/g	22–96	12	-	-	4	-	Schjorring et al., 2017
2018	Germany	RTE Meat	Sausage (cooked)	Processing contamination	-	-	-	33	-	-	5	-	JEMRA call for data
2018*	Hungary Austria, Denmark, Finland, Sweden and United Kingdom	RTE Fruit and Veg.	Frozen vegetables - corn	Processing contamination	4b	<10–150 cfu/g	-	47	-	-	10	-	ECDC and ESFA, 2018
2018*	South Africa	RTE Meat	Deli meat - Polony	Processing contamination	4b	-	<0–93 (19)	1060	455	-	216	-	Smith et al., 2019
2018	United States of America	RTE Meat	Pork products	Processing contamination	-	-	35–84 (60)	4	-	-	0	-	CDC, 2019b
2018	United States of America	RTE Meat	Deli meats - ham (sliced)	Processing contamination	-	-	70–81 (76)	4	-	-	1	-	CDC, 2018a
2017*	Italy	RTE Dairy	Cheese - soft	-	1/2a	-	-	31	-	-	-	-	Amato et al., 2017
2017	Netherlands	RTE Fish and Seafood	-	-	1/2a	-	-	13	-	-	-	-	Friesema et al., 2020

(cont.)

Year *Multiyear outbreak	Country	Food category	Subcategory	Contamination source	Serotype	Enumeration results (food products)	Age range (median)	Total no. cases (invasive) H = Hospital outbreak	Cases Maternofoetal	Immuno -comp	No. of deaths[a]	No. perinatal deaths	Ref.
2017	United States of America	RTE Dairy	Cheese – soft (unpasteurized)	Processing contamination	-	-	<1-89 (52)	10	-	-	2	-	USFDA, 2017
2017	United States of America	RTE Fruit and Veg.	Processed fruit – caramel apples	Undetermined	-	-	55-71 (69)	3	-	-	0	-	Marus et al., 2019
2016	Canada	RTE Dairy	Milk – flavored, chocolate (pasteurized)	Processing contamination - post- pasteurization pump	-	-	<1-90 (73)	34	1	-	4	-	Hanson et al., 2019
2016*	Czechia	RTE Meat	Deli meats – turkey (sliced)	Processing contamination	1/2a	36 000 -84 000 cfu/g	-	26	-	-	3	-	Gelbicova et al., 2018
2016	Germany	RTE Meat	Juniper ham	Processing contamination	-	-	-	11	-	-	1	-	JEMRA call for data
2016	Italy	RTE meat	Deli meat – Hog head cheese	Processing contamination	1/2a	-	0-88 (73)	24	1	24	4	-	Duranti et al., 2018
2016	Switzerland	RTE Meat	Pâté	Processing contamination - equipment (mincing)	4b	-	-	5	-	-	-	-	Althaus et al., 2017
2016*	United States of America	RTE Fruit and Veg.	Frozen vegetables – various	Processing contamination	-	-	56-91 (76)	9	-	-	3	-	USFDA, 2016
2016*	United States of America, Canada	RTE Fruit and Veg.	Pre-packaged leafy greens – (chopped)	Processing contamination - chopping	4b	-	3-83 (64)	33	1	-	4	0	Self et al., 2016, 2019

(cont.)

Year *Multiyear outbreak	Country	Food category	Subcategory	Contamination source	Serotype	Enumeration results (food products)	Age range (median)	Total no. cases (invasive) H = Hospital outbreak	Cases Maternofoetal	Immuno -comp	No. of deaths[a]	No. perinatal deaths	Ref.
2015*	Denmark	RTE Fish and Seafood	Salmon (cold smoked or gravad)	Processing contamination	1/2a	-	12-89 (69)	10	1	-	4	1	Lassen et al., 2016
2015*	Denmark	RTE Fish and Seafood	Halibut and trout (cold smoked)	Processing contamination	4b	-	43-90 (73)	10	0	-	3	0	Lassen et al., 2016
2015	Finland	Multiple Foods	Unknown	Unprocessed contaminated ingredients, storage time/temperature abuse	-	-	-	24	-	-	0	-	Ricci et al., 2018
2015	Germany	Multiple Foods	Rice pudding	Processing contamination - cooling	4b	-	-	159	-	-	0	-	Ricci et al., 2018
2015	Italy	RTE Meat	Pork	Unprocessed contaminated ingredients, cross contamination	1/2a	-	-	12	-	-	2	-	Ricci et al., 2018
2015	Netherlands	RTE Fish and Seafood	Salmon (smoked)	-	-	-	-	3	-	-	-	-	Friesema et al., 2020
2015	Portugal	Multiple Foods	-	Cross contamination (food service)	4b	-	-	3H	-	-	0	-	Ricci et al., 2018
2015	Sweden	Multiple Foods	Likely dill which then contaminated crustaceans and cheese	-	-	-	-	13	-	-	-	-	Ricci et al., 2018

(cont.)

Year *Multiyear outbreak	Country	Food category	Subcategory	Contamination source	Serotype	Enumeration results (food products)	Age range (median)	Total no. cases (invasive) H = Hospital outbreak	Cases Maternofoetal	Immuno -comp	No. of deaths[a]	No. perinatal deaths	Ref.
2015	United States of America	RTE Dairy	Cheese – various varieties (feta, etc.)	Processing contamination	-	-	1-92 (73)	30	6	-	3	1	CDC, 2015
2015*	United States of America	RTE Dairy	Ice cream – various	Processing contamination	1/2b, 3b, 1/2a	<20 MPN/g	-	10H	-	-	3	-	Buchanan et al., 2017; Chen et al., 2016b
2015	United States of America	RTE Dairy	Other sour – cream	-	-	-	-	2	-	-	0	-	CDC, 2018b
2014*	Denmark	RTE Meat	Deli meat – spiced meat roll (sliced)	Processing contamination	1/2b	<10-115 cfu/g	43-90 (71)	41	0	41	17	0	Jensen et al., 2016
2014	Denmark	Multiple Foods	Composite meal	-	-	-	-	6H	-	-	0	-	Ricci et al., 2018
2014	Denmark	RTE Fish and Seafood	Trout and halibut (smoked)	-	-	-	-	6	-	-	0	-	Ricci et al., 2018
2014	Germany	Multiple Foods	Iceberg lettuce with yogurt dressing and gouda cheese	-	1/2a	-	-	2H	-	-	0	-	Ricci et al., 2018
2014	Sweden	RTE Meat	Sausage	-	1/2a	-	-	4	-	-	-	-	Ricci et al., 2018
2014	Switzerland	RTE Fruit and Veg.	Pre-packaged leafy greens – (chopped)	Processing contamination - product-feeding belt	4b	100-1000 cfu/g	789 (74)	32	-	-	4	-	Stephan et al., 2015
2014	United Kingdom	Multiple Foods	Sandwiches	Other	4b	-	-	4H	-	-	0	-	Ricci et al., 2018

(cont.)

Year *Multiyear outbreak	Country	Food category	Subcategory	Contamination source	Serotype	Enumeration results (food products)	Age range (median)	Total no. cases (invasive) H = Hospital outbreak	Cases Maternofoetal	Immuno -comp	No. of deaths[a]	No. perinatal deaths	Ref.
2014	United States of America	RTE Dairy	Milk – flavored, chocolate (unpasteurized)	-	-	-	-	2	-	-	1	-	CDC, 2016
2014	United States of America	RTE Fruit and Veg.	Fresh sprouts – mung bean	Processing contamination – environmental and spent irrigation water	4b	-	-	5	-	-	2	-	Buchanan et al., 2017
2014	United States of America	RTE Dairy	Cheese – Mexican style, soft, quesito casero	Processing contamination – environment	-	-	-	5	3	-	1	-	CDC, 2014; Jackson et al., 2018
2014	United States of America	RTE Dairy	Cheese – Mexican style (queso fresco, other)	Processing contamination	-	-	-	8	5	-	1	-	CDC, 2014b; Zhu, Gooneratne and Hussain, 2017
2014	United States of America	RTE Fruit and Veg.	Fresh fruit – peaches, nectarines, whole	Undetermined	1/2b, 4b	<1-25 cfu/g	-	2	-	-	0	0	Buchanan et al., 2017; Chen et al., 2016a
2014	United States of America	RTE Fruit and Veg.	Fresh sprouts	-	-	-	-	2	-	-	0	-	CDC, 2018b
2014	United States of America	RTE Dairy	Ice cream – milkshake	Processing contamination – environment	-	-	-	2H	-	-	0	-	Beecher, 2015
2014	United States of America, Canada	RTE Fruit and Veg.	Processed fruit – caramel apples	Processing contamination – whole apple packaging environment	4b	-	5-92 (66)	36	11	-	7	1	Angelo et al., 2017; GC, 2015; Zhu, Gooneratne and Hussain, 2017

(cont.)

Year *Multiyear outbreak	Country	Food category	Subcategory	Contamination source	Serotype	Enumeration results (food products)	Age range (median)	Total no. cases (invasive) H = Hospital outbreak	Cases Maternofoetal	Immuno -comp	No. of deaths[a]	No. perinatal deaths	Ref.
2013	Belgium	RTE Meat	Pork	-	1/2a	-	-	2	-	-	0	-	Ricci et al., 2018
2013	Belgium	RTE Dairy	Cheese	Unprocessed contaminated ingredients	1/2b	-	-	2	-	-	0	-	Ricci et al., 2018
2013	Chile	RTE Dairy	Cheese – soft cheese (Camembert)	Undetermined	-	-	-	34	-	-	5	-	ISID, 2020
2013	France	RTE Fish and Seafood	Crustaceans, shell, molluscs and products thereof	-	-	-	-	3	-	-	0	-	Ricci et al., 2018
2013	Germany	RTE Fruit and Veg.	Vegetables and juices and other products thereof – mixed salad	Unprocessed contaminated ingredients	-	-	-	3H	-	-	1	-	Ricci et al., 2018
2013	Norway	RTE Fish and Seafood	Trout (half fermented)	-	-	-	-	3	-	-	1	-	Ricci et al., 2018
2013	Spain	RTE Meat	Fois gras	Processing contamination	1/2b	52 000 cfu/g	-	10	5	-	3[c]	-	Pérez-Trallero et al., 2014
2013	Sweden	RTE Meat	Meat and meat products	-	-	-	-	34	-	-	-	-	Ricci et al., 2018
2013*	United Kingdom	RTE Fish and Seafood	Crab meat	Processing – inadequate chilling	4	>100 cfu/g	> 55 years	3	-	3	1	-	McLauchlin, Grant and Amar, 2020
2013*	United Kingdom	RTE Meat	Beef - set in gelatin (cooked pressed)	-	1/2a	<20-100 cfu/g	> 60	5	-	-	-	-	McLauchlin, Grant and Amar, 2020
2013	United Kingdom	RTE Fish and Seafood	Crab meat	Processing contamination	4	<20->1 000 cfu/g	-	4	-	-	1	-	McLauchlin, Grant and Amar, 2020

c Listeriosis may have contributed to three deaths

(cont.)

Year *Multiyear outbreak	Country	Food category	Subcategory	Contamination source	Serotype	Enumeration results (food products)	Age range (median)	Total no. cases (invasive) H = Hospital outbreak	Cases Maternofoetal	Immuno -comp	No. of deaths[a]	No. perinatal deaths	Ref.
2013	United States of America	RTE Dairy	Cheese -French style, semi-soft	Processing contamination - environment	-	-	30-67 (55)	6	1	-	1	1	CDC, 2013; Jackson et al., 2018
2013	United States of America	RTE Dairy	Cheese-Latin style, soft	-	-	-	-	8	-	-	1	-	CDC, 2018b
2013	United States of America	Multiple Foods	Hummus	-	-	-	-	8	-	-	1	-	CDC, 2018b
2013	United States of America	Multiple Foods	Hummus	-	-	-	-	28	-	-	3	-	CDC, 2018b
2012	Australia	RTE Dairy	Brie, Camembert (pasteurized)	Processing contamination	4b, 4d, 4e	-	-	34	-	-	6	1	ADOHA, 2013
2012	Finland	RTE Meat	Deli meat - Meat Jelly (sliced)	Processing contamination	1/2a	-	57-95	20H	-	9	3	-	Jacks et al., 2016, Ricci et al., 2018
2012	New Zealand	RTE Meat	Locally produced meats	Processing contamination	-	-	-	4H	-	-	2	-	Rivas et al., 2019
2012*	Portugal	RTE Dairy	Cheese - (pasteurized)	Processing contamination - environment	4b	>100 cfu/g	(59)	30	2	-	11	-	Ferreira et al., 2018; Magalhaes et al., 2015
2012	Spain	RTE Dairy	Cheese - Latin-style, fresh (pasteurized)	Processing contamination	1/2a	>100-32 000 cfu/g	-	2	2	-	-	-	de Castro et al., 2012

(cont.)

Year *Multiyear outbreak	Country	Food category	Subcategory	Contamination source	Serotype	Enumeration results (food products)	Age range (median)	Total no. cases (invasive) H = Hospital outbreak	Cases Maternofoetal	Immuno -comp	No. of deaths[a]	No. perinatal deaths	Ref.
2012*	United Kingdom	Multiple Foods	Pork pies	Processing contamination	1/2a	< 20–10 000 cfu/g	> 55	14	–	10	1	–	McLauchlin, Grant and Amar, 2020
2012	United States of America	RTE Dairy	Cheese – Ricotta salata, sheep (pasteurized)	Processing contamination and cross contamination of other cheese products at the retail level	1/2a, 4b	< 10– 1100 000 cfu/g	–	23	9	–	5	1	Acciari et al., 2016; CDC, 2012a
2011	Belgium	RTE Dairy	Cheese – hard (pasteurized)	Processing contamination – surface contamination from equipment	1/2a	20 cfu/g	–	12H	–	–	2	–	Yde et al., 2012
2011	Finland	Multiple Foods	Bakery – sponge cake	Processing contamination	–	–	–	–	2	–	0	–	Ricci et al., 2018
2011	Switzerland	RTE Meat	Deli meat – ham (sliced)	Processing contamination – slicer	1/2a	470 and 4 800 cfu/g	–	6	–	2	–	–	Hachler et al., 2013
2011	United Kingdom	Multiple Foods	Sandwiches and prepared salads	Processing contamination and cross contamination	4	–	50–80	3H	–	–	0	–	Coetzee et al., 2011; Ricci et al., 2018
2011	United States of America	RTE Fruit and Veg.	Fresh cantaloupes – whole	Processing contamination – surface contamination from equipment	1/2a, 1/2b	–	< 1 to 96 (78)	147	7	–	33	1	Buchanan et al., 2017; CDC, 2012b; Laksanalamai et al., 2012

(cont.)

Year *Multiyear outbreak	Country	Food category	Subcategory	Contamination source	Serotype	Enumeration results (food products)	Age range (median)	Total no. cases (invasive) H = Hospital outbreak	Cases Maternofoetal (invasive)	Immuno -comp	No. of deaths[a]	No. perinatal deaths	Ref.
2011	United States of America	RTE Fruit and Veg.	Fresh leafy greens – Romain lettuce	–	–	–	–	–	–	–	–	–	Shrivastava, 2011; Zhu, Gooneratne and Hussain, 2017
2011	United States of America	RTE Dairy	Cheese – Ackawi and chive (pasteurized)	–	–	–	–	2	–	–	1	–	CDC, 2018b
2011	United States of America	RTE Dairy	Cheese – blue veined (unpasteurized)	–	–	–	–	15	–	–	1	–	CDC, 2018b
2010	Australia	RTE Fruit and Veg.	Fresh cantaloupes (Rock melons) – whole & cut	Unprocessed contaminated ingredients – surface contamination of raw fruit	1/2b, 3b, 7	–	–	9	–	–	2	–	ADOHA, 2012
2010*	Austria, Germany, Czechia	RTE Dairy	Cheese – Quargel	Processing contamination	1/2a	>100 cfu/g	57–89 (72)	34	0	–	8	–	Fretz et al., 2010
2010	Denmark	RTE Fish and Seafood	Salmon (gravad)	–	–	–	–	9	–	–	0	–	Ricci et al., 2018
2010	Germany	RTE Fish and Seafood	Herring – casserole in vegetable oil	–	4b	–	–	12	–	–	1	–	Ricci et al., 2018
2010	United Kingdom	RTE Meat	Bovine meat and products thereof	Cross contamination (food service)	1/2a	–	–	4	–	–	2	–	Ricci et al., 2018

(cont.)

Year *Multiyear outbreak	Country	Food category	Subcategory	Contamination source	Serotype	Enumeration results (food products)	Age range (median)	Total no. cases (invasive) H = Hospital outbreak	Cases Maternofoetal	Immuno-comp	No. of deaths[a]	No. perinatal deaths	Ref.
2010*	United Kingdom	RTE Meat	Deli meat – ham and tongue (sliced)	Processing contamination	1/2a	100-10 000 cfu/g	48-81	10	-	-	2	-	McLauchlin, Grant and Amar, 2020; Ricci et al., 2018
2010	United Kingdom	Multiple Foods	Sandwiches	Cross contamination storage time/temperature abuse; unprocessed contaminated ingredients	4	-	-	4H	-	-	1	-	Ricci et al., 2018
2010	United States of America	RTE Meat	Deli meat – hog head cheese	Processing contamination	1/2a	-	38-93 (64)	14	-	6	2	-	CDC, 2011
2010	United States of America	RTE Fruit and Veg.	Processed vegetable – Celery (chopped)	Processing contamination-equipment (slicer) and environment	1/2a	-	56-93	10H	-	10	5	-	Buchanan et al., 2017; Knudson Gaul et al., 2012
2010	United States of America	Multiple Foods	Sushi	-	-	-	-	2	-	-	0	0	CDC, 2018b
2009	Australia	RTE Meat	Deli meat – chicken (sliced)	Processing contamination	1/2c	-	-	35[d]	8[e]	-	4	3	ADOHA, 2009; Popovic, Heron and Covacin, 2014
2009	Australia	RTE Dairy	Cheese	Processing contamination	1/2a	-	-	25	-	-	5	-	Ricci et al., 2018
2009	Chile	RTE Meat	Sausages and other meat products	-	-	-	-	73	-	-	17	-	Montero et al., 2015

[d] 13 cases were lab confirmed and 22 cases were clinical.
[e] 8 of 13 lab confirmed cases were maternofoetal cases.

(cont.)

Year *Multiyear outbreak	Country	Food category	Subcategory	Contamination source	Serotype	Enumeration results (food products)	Age range (median)	Total no. cases (invasive) H = Hospital outbreak	Cases Maternofoetal	Immuno-comp	No. of deaths[a]	No. perinatal deaths	Ref.
2009	Czechia	RTE Meat	Pork – ham (vacuum packed)	Processing contamination	1/2a	-	-	9H	-	-	4	-	Gelbicova et al., 2018
2009	Denmark	RTE Meat	Beef (sliced)	Processing contamination	-	<100 cfu/g	44-94 (78)	8	0	8	2	0	Smith et al., 2011
2009	Germany	RTE Dairy	Cheese (pasteurized)	-	1/2a	-	-	6	-	-	2	-	Ricci et al., 2018
2009	United Kingdom	RTE Meat	Deli meat (sliced)	Processing contamination	4	20-10 000 cfu/g	-	14	1	11	-	-	McLauchlin, Grant and Amar, 2020
2008	Austria	RTE Meat	Deli meat – jellied pork	Post processing contamination (food service)	4b	3 000-30 000 cfu/g	-	14	-	-	-	-	Pichler et al., 2009
2008	Canada	RTE Meat	Deli meat – various (sliced)	Processing contamination - slicer	1/2a	<50->20 000 cfu/g	29-98 (78)	57H	0	41	24	-	Currie et al., 2015
2008	Canada	RTE Dairy	Cheese (pasteurized)	Processing contamination - environment	1/2a	>104 cfu/g	28-89 (65)	38	16	21	-	3	Gaulin, Ramsay and Bekal, 2012
2008	Chile	RTE Dairy	Cheese – brie and camembert, soft	-	-	-	-	165	-	-	14	-	Montero et al., 2015
2008*	Germany	RTE Meat	Scalded sausages (sliced)	Processing contamination - slicer	4b	<10-115 cfu/g	(67)	16[i]	-	16	5	-	Winter et al., 2015

[i] Nine cases were lab confirmed and seven were probable cases.

(cont.)

Year * Multiyear outbreak	Country	Food category	Subcategory	Contamination source	Serotype	Enumeration results (food products)	Age range (median)	Total no. cases (invasive) H = Hospital outbreak	Cases Maternofoetal	Immuno -comp	No. of deaths[a]	No. perinatal deaths	Ref.
2008*	United States of America	RTE Dairy	Cheese - Mexican style (pasteurized)	Processing contamination - vat gasket in a post pasteurization	-	-	3-43 (31)	8	7	-	0	2	Jackson et al., 2011
2008	United States of America	Multiple Foods	Tuna salad	-	1/2a	-	-	5H	-	5	3	-	Cartwright et al., 2013; Cokes et al., 2011
2008	United States of America	RTE Fruit and Veg.	Fresh sprouts	-	1/2a	-	-	20	4	-	0	-	Cartwright et al., 2013
2007*	Germany	RTE Dairy	Cheese (pasteurized)	Processing contamination - environment	4b	52 000 - 120 000 cfu/g	(69)	34H[g]	11	-	-	-	Koch et al., 2010
2007	Norway	RTE Dairy	Cheese - Camembert (pasteurized)	Processing contamination - environment	-	360 million cfu/g	27-84 (64)	17H	-	15	3	-	Johnsen et al., 2010
2007	United States of America	RTE Dairy	Milk – white and flavored (pasteurized)	Processing contamination - environment	4b	-	31-87 (75)	5	2	-	3	1	Cartwright et al., 2013
2006	Czechia	RTE Dairy	Cheese	Processing contamination	1/2b	-	-	26	-	-	-	-	Vit et al., 2007
2006	United States of America	Multiple Foods	Taco or nacho salad	-	1/2b	-	-	2	-	-	0	-	Cartwright et al., 2013
2006	United States of America	RTE Dairy	Cheese (pasteurized)	-	4b	-	-	3	-	-	1	-	Cartwright et al., 2013

[a] Estimated number of cases based on serotyping and cheese exposure information.

(cont.)

Year * Multiyear outbreak	Country	Food category	Subcategory	Contamination source	Serotype	Enumeration results (food products)	Age range (median)	Total no. cases (invasive) H = Hospital outbreak	Cases Maternofoetal	Immuno -comp	No. of deaths[a]	No. perinatal deaths	Ref.
2005	Australia	RTE Meat	Deli meat - corned beef	Processing contamination - slicer	-	-	-	4H	-	-	2	-	Popovic, Heron and Covacin, 2014
2005	Switzerland	RTE Dairy	Cheese - Tomme, soft	Processing contamination - environment	1/2a	Cheese: 1 000-10 000 cfu/g Butter: 10-100 cfu/g	-	10	2	8	3	2	Bille et al., 2006
2005	United States of America	RTE Meat	Chicken (grilled, sliced)	Processing contamination - slicer	1/2b	-	-	3	-	-	0	-	Cartwright et al., 2013
2005	United States of America	RTE Meat	Deli meat - turkey	-	1/2a	-	-	37	-	-	1	-	Cartwright et al., 2013
2005	United States of America	RTE Dairy	Cheese - Mexican style (unpasteurized)	-	4b	-	-	36	-	-	0	-	Cartwright et al., 2013

A2.5 REFERENCES

Acciari, V.A., Iannetti, L., Gattuso, A., Sonnessa, M., Scavia, G., Montagna, C., Addante, N., Torresi, M., Zocchi, L., Scattolini, S., Centorame, P., Marfoglia, C., Prencipe, V.A. & Gianfranceschi, M.V. 2016. Tracing sources of *Listeria* contamination in traditional Italian cheese associated with a US outbreak: investigations in Italy. *Epidemiology and Infection*, 144(13): 2719–2727. https://doi.org/10.1017/s095026881500254x

Althaus, D., Jermini, M., Giannini, P., Martinetti, G., Reinholz, D., Nüesch-Inderbinen, M., Lehner, A. & Stephan, R. 2017. Local outbreak of *Listeria monocytogenes* serotype 4b sequence type 6 due to contaminated meat Pâté. *Foodborne Pathogens and Disease*, 14(4): 219–222. https://doi.org/10.1089/fpd.2016.2232

Amato, E., Filipello, V., Gori, M., Lomonaco, S., Losio, M.N., Parisi, A., Huedo, P., Knabel, S.J. & Pontello, M. 2017. Identification of a major *Listeria monocytogenes* outbreak clone linked to soft cheese in Northern Italy - 2009-2011. *BMC Infectious Diseases*, 17(1): 342. https://doi.org/10.1186/s12879-017-2441-6

Angelo, K.M., Conrad, A.R., Saupe, A., Dragoo, H., West, N., Sorenson, A., Barnes, A., Doyle, M., Beal, J., Jackson, K.A., Stroika, S., Tarr, C., Kucerova, Z., Lance, S., Gould, L.H., Wise, M. & Jackson, B.R. 2017. Multistate outbreak of *Listeria monocytogenes* infections linked to whole apples used in commercially produced, prepackaged caramel apples: United States, 2014-2015. *Epidemiology and Infection*, 145(5): 848–856. https://doi.org/10.1017/S0950268816003083

ADOHA (Australian Government Department of Health and Ageing). 2009. Department of Health | OzFoodNet quarterly report, 1 July to 30 September 2009. In: *Australian Government Department of Health*. Canberra ACT, Australia. Cited 5 March 2020. www1.health.gov.au/internet/main/publishing.nsf/Content/cda-cdi3304f.htm

ADOHA. 2012. Monitoring the incidence and causes of diseases potentially transmitted by food in Australia: annual report of the OzFoodNet network, 2010. *Communicable Diseases Intelligence Quarterly Report*, 36(3): E213–41. www.ncbi.nlm.nih.gov/pubmed/23186234

ADOHA. 2013. OzFoodNet quarterly report, 1 October to 31 December 2012. *Communicable Diseases Intelligence Quarterly Report*, 37(4). www1.health.gov.au/internet/main/publishing.nsf/Content/cda-cdi3704-pdf-cnt.htm/$FILE/cdi3704g.pdf

ADOHA. 2019. *Listeria* cases among at-risk persons are a timely reminder for food safety | Australian Government Department of Health. In: *Australian Government Department of Health*. Canberra ACT, Australia. Cited 5 March 2020. www.health.gov.au/news/listeria-cases-among-at-risk-persons-are-a-timely-reminder-for-food-safety-0

BBC News. 2019. "Smoked salmon" listeria kills two in Australia. In: *BBC News*. London, UK. Cited 5 March 2020. www.bbc.com/news/world-australia-49094165

Beach, C. 2020. Fatalities reported in *Listeria* outbreak traced to imported mushrooms. In: *Food Safety News*. Seattle, WA., USA. Cited 18 March 2020. www.foodsafetynews.com/2020/03/fatalities-reported-in-listeria-outbreak-traced-to-imported-mushrooms/

Beecher, C. 2015. Lesson from WA ice cream recall: don't let it happen to you. In: *Food Safety News*. Seattle, WA., USA. Cited 6 March 2020. www.foodsafetynews.com/2015/01/dont-let-a-recall-happen-to-you-warns-owner-of-ice-cream-company/

Bille, J., Blanc, D.S., Schmid, H., Boubaker, K., Baumgartner, A., Siegrist, H.H., Tritten, M.L., Lienhard, R., Berner, D., Anderau, R., Treboux, M., Ducommun, J.M., Malinverni, R., Genné, D., Erard, P.H. & Waespi, U. 2006. Outbreak of human listeriosis associated with tomme cheese in northwest Switzerland, 2005. *Eurosurveillance*, 11(6): 91–93. https://www.mendeley.com/catalogue/238ea84d-f3db-3160-b690-0bda796cac02/

Buchanan, R.L., Gorris, L.G.M., Hayman, M.M., Jackson, T.C. & Whiting, R.C. 2017. A review of *Listeria monocytogenes*: An update on outbreaks, virulence, dose-response, ecology, and risk assessments. *Food Control*, 75: 1–13.

Cabal, A., Allerberger, F., Huhulescu, S., Kornschober, C., Springer, B., Schlagenhaufen, C., Wassermann-Neuhold, M., Fötschl, H., Pless, P., Krause, R., Lennkh, A., Murer, A., Ruppitsch, W. & Pietzka, A. 2019. Listeriosis outbreak likely due to contaminated liver pâté consumed in a tavern, Austria, December 2018. *Eurosurveillance*, 24(39): 1900274. https://doi.org/10.2807/1560-7917.ES.2019.24.39.1900274

Cartwright, E.J., Jackson, K.A., Johnson, S.D., Graves, L.M., Silk, B.J. & Mahon, B.E. 2013. Listeriosis outbreaks and associated food cehicles, United States, 1998-2008. *Emerging Infectious Diseases*, 19(1): 1–9. https://doi.org/10.3201/eid1901.120393

de Castro, V., Escudero, J.M., Rodriguez, J.L., Muniozguren, N., Uribarri, J., Saez, D. & Vazquez, J. 2012. Listeriosis outbreak caused by Latin-style fresh cheese, Bizkaia, Spain, August 2012. *Eurosurveillance*, 17(42): 8–10.

CDC. 2011. Outbreak of invasive listeriosis associated with the consumption of hog head cheese - Louisiana, 2010. *Morbidity and Mortality Weekly Report*, 60(13): 401–405. www.cdc.gov/mmwr/preview/mmwrhtml/mm6013a2.htm

CDC. 2012a. Multistate outbreak of listeriosis linked to imported rescolina Marte Brand Ricotta Salata cheese (Final Update). In: *Centers for Disease Control and Prevention*. Atlanta, Georgia, USA. Cited 22 April 2022. www.cdc.gov/listeria/outbreaks/cheese-09-12/

CDC. 2012b. Multistate outbreak of listeriosis linked to whole Cantaloupes from Jensen Farms, Colorado (FINAL UPDATE). In: *Centers for Disease Control and Prevention*. Atlanta, Georgia, USA. Cited 22 April 2022. www.cdc.gov/listeria/outbreaks/cantaloupes-jensen-farms/index.html

CDC. 2013. Multistate outbreak of listeriosis linked to Crave Brothers Farmstead cheeses. In: *Centers for Disease Control and Prevention*. Atlanta, Georgia, USA. Cited 6 March 2020. www.cdc.gov/listeria/outbreaks/cheese-07-13/index.html

CDC. 2014a. Oasis Brands, Inc. Cheese recalls and investigation of human listeriosis cases. In: *Centers for Disease Control and Prevention*. Atlanta, Georgia, USA. Cited 5 March 2020. www.cdc.gov/listeria/outbreaks/cheese-10-14/index.html

CDC. 2014b. Multistate outbreak of listeriosis linked to Roos Foods Dairy Products (Final Update). In: *Centers for Disease Control and Prevention*. Atlanta, Georgia, USA. Cited 22 April 2022 www.cdc.gov/listeria/outbreaks/cheese-02-14/index.html

CDC. 2015. Multistate outbreak of listeriosis linked to soft cheeses distributed by Karoun Dairies, Inc. (Final Update). In: *Centers for Disease Control and Prevention*. Atlanta, Georgia, USA. Cited 22 April 2022 www.cdc.gov/listeria/outbreaks/soft-cheeses-09-15/index.html

CDC. 2016. Multistate outbreak of listeriosis linked to raw milk produced by Miller's Organic Farm in Pennsylvania (Final Update). In: *Centers for Disease Control and Prevention*. Atlanta, Georgia, USA. Cited 22 April 2022 www.cdc.gov/listeria/outbreaks/raw-milk-03-16/index.html

CDC. 2018a. Outbreak of *Listeria* infections linked to Deli ham (Final Update). In: *Centers for Disease Control and Prevention*. Atlanta, Georgia, USA. Cited 22 April 2022 www.cdc.gov/listeria/outbreaks/countryham-10-18/index.html

CDC. 2018b. National Outbreak Reporting System (NORS). In: *Centers for Disease Control and Prevention*. Atlanta, Georgia, USA. Cited 22 April 2022 www.cdc.gov/norsdashboard/

CDC. 2019a. Outbreak of *Listeria* infections linked to Deli-sliced meats and cheeses. In: *Centers for Disease Control and Prevention*. Atlanta, Georgia, USA. Cited 6 March 2020. www.cdc.gov/listeria/outbreaks/deliproducts-04-19/index.html)

CDC. 2019b. Outbreak of Listeria infections linked to pork products. In: *Centers for Disease Control and Prevention*. Atlanta, Georgia, USA. Cited 22 April 2022 www.cdc.gov/listeria/outbreaks/porkproducts-11-18/index.html

Chailloux, M. 2020. Cantal : la "Ferme de Gioux" fermée administrativement après des cas de listeriose. In: *France Bleu, Pays d'Auvergne*. Cited 12 October 2020. www.francebleu.fr/infos/agriculture-peche/cantal-la-ferme-de-gioux-fermee-administrativement-apres-des-cas-de-listeriose-1583922302

Chen, Y., Burall, L.S., Luo, Y., Timme, R., Melka, D., Muruvanda, T., Payne, J., Wang, C., Kastanis, G., Maounounen-Laasri, A., De Jesus, A.J., Curry, P.E., Stones, R., K'Aluoch, O., Liu, E., Salter, M., Hammack, T.S., Evans, P.S., Parish, M., Allard, M.W., Datta, A., Strain, E.A. & Brown, E.W. 2016a. *Listeria monocytogenes* in stone fruits linked to a multistate outbreak: enumeration of cells and whole-genome sequencing. *Applied and Environmental Microbiology*, 82(24): 7030–7040. https://doi.org/10.1128/AEM.01486-16

Chen, Y.I., Burall, L.S., Macarisin, D., Pouillot, R., Strain, E., De Jesus, A.J., Laasri, A., Wang, H., Ali, L., Tatavarthy, A., Zhang, G., Hu, L., Day, J., Kang, J., Sahu, S., Srinivasan, D., Klontz, K., Parish, M., Evans, P.S., Brown, E. W., Hammack, T.S., Zink, D. & Datta, A.R. 2016b. Prevalence and level of *Listeria monocytogenes* in ice cream linked to a listeriosis outbreak in the United States. *Journal of Food Protection*, 79(11): 1828–1832. https://doi.org/10.4315/0362-028X.JFP-16-208

Coetzee, N., Laza-Stanca, V., Orendi, J.M., Harvey, S., Elviss, N.C. & Grant, K.A. 2011. A cluster of *Listeria monocytogenes* infections in hospitalised adults, Midlands, England, February 2011. *Eurosurveillance*, 16(20): 19869. www.eurosurveillance.org/content/10.2807/ese.16.20.19869-en

Cokes, C., France, A.M., Reddy, V., Hanson, H., Lee, L., Kornstein, L., Stavinsky, F. & Balter, S. 2011. Serving high-risk foods in a high-risk setting: survey of hospital food service practices after an outbreak of listeriosis in a hospital. *Infection Control and Hospital Epidemiology*, 32(4): 380–386. https://doi.org/10.1086/658943

Currie, A., Farber, J.M., Nadon, C., Sharma, D., Whitfield, Y., Gaulin, C., Galanis, E., Bekal, S., Flint, J., Tschetter, L., Pagotto, F., Lee, B., Jamieson, F., Badiani, T., MacDonald, D., Ellis, A., May-Hadford, J., McCormick, R., Savelli, C., Middleton, D., Allen, V., Tremblay, F.W., MacDougall, L., Hoang, L., Shyng, S., Everett, D., Chui, L., Louie, M., Bangura, H., Levett, P.N., Wilkinson, K., Wylie, J., Reid, J., Major, B., Engel, D., Douey, D., Huszczynski, G., Di Lecci, J., Strazds, J., Rousseau, J., Ma, K., Isaac, L., Sierpinska, U. & Natl Outbreak Invest, T. 2015. Multi-province listeriosis outbreak linked to contaminated Deli meat consumed primarily in institutional settings, Canada, 2008. *Foodborne Pathogens and Disease*, 12(8): 645–652. https://doi.org/10.1089/fpd.2015.1939

Duranti, A., Sabbatucci, M., Blasi, G., Acciari, V.A., Ancora, M., Bella, A., Busani, L., Centorame, P., Camma, C., Conti, F., De Medici, D., Di Domenico, M., Di Marzio, V., Filippini, G., Fiore, A., Fisichella, S., Gattuso, A., Gianfranceschi, M., Graziani, C., Guidi, F., Marcacci, M., Marfoglia, C., Neri, D., Orsini, M., Ottaviani, D., Petruzzelli, A., Pezzotti, P., Rizzo, C., Ruolo, A., Scavia, G., Scuota, S., Tagliavento, G., Tibaldi, A., Tonucci, F., Torresi, M., Migliorati, G. & Pomilio, F. 2018. A severe outbreak of listeriosis in central Italy with a rare pulsotype associated with processed pork products. *Journal of Medical Microbiology*, 67(9): 1351–1360. https://doi.org/10.1099/jmm.0.000785

DW News. 2019. Listeria-tainted sausage deaths in Germany lead to calls for better consumer protection. In: *DW.* Berlin, Germany. Cited 12 October 2020. www.dw.com/en/listeria-tainted-sausage-deaths-in-germany-lead-to-calls-for-better-consumer-protection/a-50711199

ECDC (European Centre for Disease Prevention and Control) and EFSA (European Food Safety Authority). 2019. *Multi-country outbreak of Listeria monocytogenes clonal complex 8 infections linked to consumption of cold-smoked fish products – 4 June 2019.* Stockholm and Parma. www.ecdc.europa.eu/sites/default/files/documents/20190423_Joint_ECDC-EFSA_ROA_UI-452_Lm-ST1247.pdf

ECDC & EFSA. 2018. *Multi-country outbreak of Listeria monocytogenes serogroup IVb, multi-locus sequence type 6, infections probably linked to frozen corn.* Stockholm and Parma. https://efsa.onlinelibrary.wiley.com/doi/epdf/10.2903/sp.efsa.2018.EN-1402

EFSA & ECDC. 2019. *Multi-country outbreak of Listeria monocytogenes sequence type 6 infections linked to ready-to-eat meat products.* Parma, Italy. www.ecdc.europa.eu/sites/default/files/documents/Listeria-rapid-outbreak-assessment-NL-BE.pdf

Ferreira, V., Magalhaes, R., Almeida, G., Cabanes, D., Fritzenwanker, M., Chakraborty, T., Hain, T. & Teixeira, P. 2018. Genome sequence of *Listeria monocytogenes* 2542, a serotype 4b strain from a cheese-related outbreak in Portugal. *Microbiology Resource Announcements*, 6(25): 2. https://doi.org/10.1128/genomeA.00540-18

FSN (Food Safety News). 2020a. French authorities link *Listeria* infections to cheese company. In: *Food Safety News.* Seattle, Washington, USA. Cited 12 October 2020. www.foodsafetynews.com/2020/03/french-authorities-link-listeria-infections-to-cheese-company/#more-192952

FSN. 2020b. Cheese firm in Switzerland investigated over *Listeria* link. In: *Food Safety News.* Seattle, Washington, USA. Cited 12 October 2020. www.foodsafetynews.com/2020/08/cheese-firm-in-switzerland-investigated-over-listeria-link/

FSN. 2020c. Offerman *Listeria* recall cost Ter Beke almost $9 million. In: *Food Safety News.* Seattle, Washington, USA. Cited 12 October 2020. www.foodsafetynews.com/2020/03/offerman-listeria-recall-cost-ter-beke-almost-9-million/

FSN. 2020d. Outbreaks down but illnesses up for Denmark. In: *Food Safety News.* Seattle, Washington, USA. Cited 12 October 2020. www.foodsafetynews.com/2020/09/outbreaks-down-but-illnesses-up-for-denmark/

Fretz, R., Sagel, U., Ruppitsch, W., Pietzka, A.T., Stöger, A., Huhulescu, S., Heuberger, S., Pichler, J., Much, P., Pfaff, G., Stark, K., Prager, R., Flieger, A., Feenstra, O. & Allerberger, F. 2010. Listeriosis outbreak caused by acid curd cheese "Quargel", Austria and Germany 2009. *Eurosurveillance*, 15(5): 1–2. https://doi.org/10.2807/ese.15.05.19477-en

Friesema, I.H.M., Slegers-Fitz-James, I.A., Wit, B. & E., F. 2020. *Voedselgerelateerde uitbraken in Nederland 2006-2017*. Rijksinstituut voor Volksgezondheid en Milieu, https://rivm.openrepository.com/handle/10029/623691

FSA (Food Standards Agency). 2019. Update on investigation into food supply chain linked to listeria. In: *Food Standards Agency*. London, UK. Cited 22 April 2022. www.food.gov.uk/news-alerts/news/update-on-investigation-into-food-supply-chain-linked-to-listeria

Gaulin, C., Ramsay, D. & Bekal, S. 2012. Widespread listeriosis outbreak attributable to pasteurized cheese, which led to extensive cross-contamination affecting cheese retailers, Quebec, Canada, 2008. *Journal of Food Protection*, 75(1): 71–78. https://doi.org/10.4315/0362-028x.jfp-11-236

Gelbicova, T., Zobanikova, M., Tomastikova, Z., Van Walle, I., Ruppitsch, W. & Karpiskova, R. 2018. An outbreak of listeriosis linked to turkey meat products in the Czech Republic, 2012-2016. *Epidemiology and Infection*, 146(11): 1407–412. https://doi.org/10.1017/s0950268818001565

GC (Government of Canada). 2015. Public Health Notice Update – United States Outbreak of *Listeria* infections related to prepackaged caramel apples that may have been distributed in Canada - Canada.ca. In: *Government of Canada*. Ottawa, ON, Canada. Cited 12 October 2020. www.canada.ca/en/public-health/services/public-health-notices/2014/public-health-notice-update-united-states-outbreak-listeria-infections-related-prepackaged-caramel-apples-that-may-have-been-distributed-canada.html

Hachler, H., Marti, G., Giannini, P., Lehner, A., Jost, M., Beck, J., Weiss, F., Bally, B., Jermini, M., Stephan, R. & Baumgartner, A. 2013. Outbreak of listerosis due to imported cooked ham, Switzerland 2011. *Eurosurveillance*, 18(18): 7–13.

Halbedel, S., Wilking, H., Holzer, A., Kleta, S., Fischer, M.A., Lüth, S., Pietzka, A., Huhulescu, S., Lachmann, R., Krings, A., Ruppitsch, W., Leclercq, A., Kamphausen, R., Meincke, M., Wagner-Wiening, C., Contzen, M., Kraemer, I.B., Al Dahouk, S., Allerberger, F., Stark, K. & Flieger, A. 2020. Large nationwide outbreak of invasive listeriosis associated with blood sausage, Germany, 2018-2019. *Emerging Infectious Diseases*, 26(7): 1456–1464. https://doi.org/10.3201/eid2607.200225

Hanson, H., Whitfield, Y., Lee, C., Badiani, T., Minielly, C., Fenik, J., Makrostergios, T., Kopko, C., Majury, A., Hillyer, E., Fortuna, L., Maki, A., Murphy, A., Lombos, M., Zittermann, S., Yu, Y., Hill, K., Kong, A., Sharma, D. & Warshawsky, B. 2019. *Listeria monocytogenes* associated with pasteurized chocolate milk, Ontario, Canada. *Emerging Infectious Diseases*, 25(3): 581–584. https://doi.org/10.3201/eid2503.180742

ISID (International Society For Infectious Disease). 2020. ProMED-mail. In: *ProMed.* Cited 12 October 2020. https://promedmail.org/promed-posts/

Jacks, A., Pihlajasaari, A., Vahe, M., Myntti, A., Kaukoranta, S.S., Elomaa, N., Salmenlinna, S., Rantala, L., Lahti, K., Huusko, S., Kuusi, M., Siitonen, A. & Rimhanen-Finne, R. 2016. Outbreak of hospital-acquired gastroenteritis and invasive infection caused by *Listeria monocytogenes*, Finland, 2012. *Epidemiology and Infection*, 144(13): 2732–2742. https://doi.org/10.1017/s0950268815002563

Jackson, K.A., Biggerstaff, M., Tobin-D'Angelo, M., Sweat, D., Klos, R., Nosari, J., Garrison, O., Boothe, E., Saathoff-Huber, L., Hainstock, L. & Fagan, R.P. 2011. Multistate outbreak of *Listeria monocytogenes* associated with Mexican-style cheese made from pasteurized milk among pregnant, hispanic women. *Journal of Food Protection*, 74(6): 949–953. https://doi.org/10.4315/0362-028x.jfp-10-536

Jackson, K.A., Gould, L.H., Hunter, J.C., Kucerova, Z. & Jackson, B. 2018. Listeriosis outbreaks associated with soft cheeses, United States, 1998–2014. *Emerging Infectious Diseases*, 24(6): 1116–1118. https://doi.org/10.3201/eid2406.171051

Jensen, A.K., Nielsen, E.M., Bjorkman, J.T., Jensen, T., Muller, L., Persson, S., Bjerager, G., Perge, A., Krause, T.G., Kiil, K., Sorensen, G., Andersen, J.K., Molbak, K. & Ethelberg, S. 2016. Whole-genome sequencing used to investigate a nationwide outbreak of listeriosis caused by ready-to-eat Delicatessen meat, Denmark, 2014. *Clinical Infectious Diseases*, 63(1): 64–78. https://doi.org/10.1093/cid/ciw192

Johnsen, B.O., Lingaas, E., Torfoss, D., Strom, E.H. & Nordoy, I. 2010. A large outbreak of *Listeria monocytogenes* infection with short incubation period in a tertiary care hospital. *Journal of Infection*, 61(6): 465–470. https://doi.org/10.1016/j.jinf.2010.08.007

Knudson Gaul, L., Farag, N.H., Shim, T., Kingsley, M.A., Silk, B.J. & Hyytia-Trees, E. 2012. Hospital-acquired listeriosis outbreak caused by contaminated diced celery-Texas, 2010. *Clinical Infectious Diseases*, 56(1): 20–26. https://doi.org/10.1093/cid/cis817

Koch, J., Dworak, R., Prager, R., Becker, B., Brockmann, S., Wicke, A., Wichmann-Schauer, H., Hof, H., Werber, D. & Stark, K. 2010. Large listeriosis outbreak linked to cheese made from pasteurized milk, Germany, 2006-2007. *Foodborne Pathogens and Disease*, 7(12): 1581–1584. https://doi.org/10.1089/fpd.2010.0631

Laksanalamai, P., Joseph, L.A., Silk, B.J., Burall, L.S., Tarr, C.L., Gerner-Smidt, P. & Datta, A.R. 2012. Genomic characterization of *Listeria monocytogenes* strains involved in a multistate listeriosis outbreak associated with cantaloupe in US. *PLoS One*, 7(7): e42448. https://doi.org/10.1371/journal.pone.0042448

Lassen, S.G., Ethelberg, S., Bjorkman, J.T., Jensen, T., Sorensen, G., Jensen, A.K., Muller, L., Nielsen, E.M. & Molbak, K. 2016. Two listeria outbreaks caused by smoked fish consumption-using whole-genome sequencing for outbreak investigations. *Clinical Microbiology and Infection*, 22(7): 620–624. https://doi.org/10.1016/j.cmi.2016.04.017

Magalhaes, R., Almeida, G., Ferreira, V., Santos, I., Silva, J., Mendes, M.M., Pita, J., Mariano, G., Mancio, I., Sousa, M.M., Farber, J., Pagotto, F. & Teixeira, P. 2015. Cheese-related listeriosis outbreak, Portugal, March 2009 to February 2012. *Euro surveillance*, 20(17): 14–19. https://doi.org/10.2807/1560-7917.es2015.20.17.21104

Marus, J.R., Bidol, S., Altman, S.M., Oni, O., Parker-Strobe, N., Otto, M., Pereira, E., Buchholz, A., Huffman, J., Conrad, A.R. & Wise, M.E. 2019. Notes from the field: outbreak of listeriosis likely associated with prepackaged caramel apples — United States, 2017. *Morbidity and Mortality Weekly Report*, 68(3): 76–77. www.cdc.gov/mmwr/volumes/68/wr/mm6803a5.htm

McLauchlin, J., Grant, K.A. & Amar, C.F.L. 2020. Human foodborne listeriosis in England and Wales, 1981 to 2015. *Epidemiology and Infection*, 148: e54. https://doi.org/10.1017/S0950268820000473

Montero, D., Bodero, M., Riveros, G., Lapierre, L., Gaggero, A., Vidal, R.M. & Vidal, M. 2015. Molecular epidemiology and genetic diversity of *Listeria monocytogenes* isolates from a wide variety of ready-to-eat foods and their relationship to clinical strains from listeriosis outbreaks in Chile. *Frontiers in Microbiology*, 6: 384. https://doi.org/10.3389/fmicb.2015.00384

NSW Department of Primary Industries (NSW DPI). 2018. *Listeria* outbreak investigation - summary report for the melon industry, October 18. In: *NSW Government Department of Primary Industries*. Cited 6 March 2020.www.foodauthority.nsw.gov.au/sites/default/files/_Documents/foodsafetyandyou/listeria_outbreak_investigation.pdf

Pérez-Trallero, E., Zigorraga, C., Artieda, J., Alkorta, M. & Marimon, J.M. 2014. Two outbreaks of *Listeria monocytogenes* infection, Northern Spain. *Emerging Infectious Diseases*, 20(12): 2155–2157. https://doi.org/10.3201/eid2012.140993

Pichler, J., Much, P., Kasper, S., Fretz, R., Auer, B., Kathan, J., Mann, M., Huhulescu, S., Ruppitsch, W., Pietzka, A., Silberbauer, K., Neumann, C., Gschiel, E., de Martin, A., Schuetz, A., Gindl, J., Neugschwandtner, E. & Allerberger, F. 2009. An outbreak of febrile gastroenteritis associated with jellied pork contaminated with *Listeria monocytogenes*. *Wiener Klinische Wochenschrift*, 121(3–4): 149–156. https://doi.org/10.1007/s00508-009-1137-3

Popovic, I., Heron, B. & Covacin, C. 2014. *Listeria*: an Australian perspective (2001-2010). *Foodborne Pathogens and Disease*, 11(6): 425–432. https://doi.org/10.1089/fpd.2013.1697

Public Health Agency of Canada (PHAC). 2019. Public health notice - outbreak of *Listeria* infections linked to Rosemount brand cooked diced chicken. In: *Government of Canada*. Ottawa, ON, Canada. Cited 12 July 2022. https://www.canada.ca/en/public-health/services/public-health-notices/2019/outbreak-listeria-infections-cooked-diced-chicken.html

Public Health England (PHE). 2019. News story - *Listeria* cases being investigated. In: *Gov.UK*. London, UK. Cited insert date. www.gov.uk/government/news/listeria-cases-being-investigated#history

Ricci, A., Allende, A., Bolton, D., Chemaly, M., Davies, R., Fernández Escámez, P.S., Girones, R., Herman, L., Koutsoumanis, K., Nørrung, B., Robertson, L., Ru, G., Sanaa, M., Simmons, M., Skandamis, P., Snary, E., Speybroeck, N., Ter Kuile, B., Threlfall, J., Wahlström, H., Takkinen, J., Wagner, M., Arcella, D., Da Silva Felicio, M.T., Georgiadis, M., Messens, W. & Lindqvist, R. 2018. *Listeria monocytogenes* contamination of ready-to-eat foods and the risk for human health in the EU. *EFSA Journal*, 16(1): 5134. https://doi.org/10.2903/j.efsa.2018.5134

Rivas, L., Dupont, P.Y., Wilson, M., Rohleder, M. & Gilpin, B. 2019. An outbreak of multiple genotypes of *Listeria monocytogenes* in New Zealand linked to contaminated ready-to-eat meats-a retrospective analysis using whole-genome sequencing. *Letters in Applied Microbiology*, 69(6): 392–398. https://doi.org/10.1111/lam.13227

Schjørring, S., Lassen, S.G., Jensen, T., Moura, A., Kjeldgaard, J.S., Mueller, L., Thielke, S., Leclercq, A., Maury, M.M., Tourdjman, M., Donguy, M.P., Lecuit, M., Ethelberg, S. & Nielsen, E.M. 2017. Cross-border outbreak of listeriosis caused by cold-smoked salmon, revealed by integrated surveillance and whole genome sequencing (WGS), Denmark and France, 2015 to 2017. *Eurosurveillance*, 22(50): 8–12. https://doi.org/10.2807/1560-7917.es.2017.22.50.17-00762

Self, J.L., Conrad, A., Stroika, S., Jackson, A., Burnworth, L., Beal, J., Wellman, A., Jackson, K.A., Bidol, S., Gerhardt, T., Hamel, M., Franklin, K., Kopko, C., Kirsch, P., Wise, M.E. & Basler, C. 2016. Note from the field: outbreak of listeriosis associated with consumption of packaged salad — United States and Canada, 2015-2016. *Morbidity and Mortality Weekly Report*, 65(33): 879–881. https://doi.org/10.15585/mmwr.mm6533a6

Self, J.L., Conrad, A., Stroika, S., Jackson, A., Whitlock, L., Jackson, K.A., Beal, J., Wellman, A., Fatica, M.K., Bidol, S., Huth, P.P., Hamel, M., Franklin, K., Tschetter, L., Kopko, C., Kirsch, P., Wise, M.E. & Basler, C. 2019. Multistate outbreak of listeriosis associated with packaged leafy green salads, United States and Canada, 2015–2016. *Emerging Infectious Diseases*, 25(8): 1461–1468. https://doi.org/10.3201/eid2508.180761

Shrivastava, S. 2011. *Listeria* outbreak - bacteria found in Romaine Lettuce: FDA. In: *International Business Times*. Cited 5 March 2020. www.ibtimes.com/listeria-outbreak-bacteria-found-romaine-lettuce-fda-320544

SMDHU (Simcoe Muskoko District Health Unit). 2018. *Listeriosis Investigation: Additional cases linked to Druxy's Famous Deli at Princess Margaret Cancer Centre.* Ontario, Canada. www.simcoemuskokahealth.org/docs/default-source/jfy-healthfax/181011listeriosis_final.pdf?sfvrsn=0

Smith, A.M., Tau, N.P., Smouse, S.L., Allam, M., Ismail, A., Ramalwa, N.R., Disenyeng, B., Ngomane, M. & Thomas, J. 2019. Outbreak of *Listeria monocytogenes* in South Africa, 2017-2018: laboratory activities and experiences associated with whole-genome sequencing analysis of isolates. *Foodborne Pathogens and Disease*, 16(7): 524–530. https://doi.org/10.1089/fpd.2018.2586

Smith, B., Larsson, J.T., Lisby, M., Muller, L., Madsen, S.B., Engberg, J., Bangsborg, J., Ethelberg, S. & Kemp, M. 2011. Outbreak of listeriosis caused by infected beef meat from a meals-on-wheels delivery in Denmark 2009. *Clinical Microbiology and Infection*, 17(1): 50–52. https://doi.org/10.1111/j.1469-0691.2010.03200.x

Stephan, R., Althaus, D., Kiefer, S., Lehner, A., Hatz, C., Schmutz, C., Jost, M., Gerber, N., Baumgartner, A., Hachler, H. & Mausezahl-Feuz, M. 2015. Foodborne transmission of *Listeria monocytogenes* via ready-to-eat salad: A nationwide outbreak in Switzerland, 2013-2014. *Food Control*, 57: 14–17. https://doi.org/10.1016/j.foodcont.2015.03.034

USFDA (United States Food & Drug Association). 2016. FDA investigated listeria outbreak linked to frozen vegetables. In: *United States Food and Drug Administration*. Washington DC. Cited 12 July 2022. www.fda.gov/food/outbreaks-foodborne-illness/fda-investigated-listeria-outbreak-linked-frozen-vegetables

USFDA. 2017. FDA investigates listeria outbreak linked to soft cheese produced by Vulto Creamery. In: *United States Food and Drug Administration*. Washington DC. Cited 12 July 2022. www.fda.gov/food/outbreaks-foodborne-illness/fda-investigates-listeria-outbreak-linked-soft-cheese-produced-vulto-creamery

USFDA. 2019. Outbreak investigation of *Listeria monocytogenes*: hard-boiled eggs (December 2019). In: *United States Food and Drug Administration*. Washington DC. Cited 12 July 2022. https://www.fda.gov/food/outbreaks-foodborne-illness/outbreak-investigation-listeria-monocytogenes-hard-boiled-eggs-december-2019

USFDA. 2020. Outbreak investigation of *Listeria monocytogenes*: enoki mushrooms (March 2020). In: *United States Food and Drug Administration*. Washington DC. Cited 12 October 2020. www.fda.gov/food/outbreaks-foodborne-illness/outbreak-investigation-listeria-monocytogenes-enoki-mushrooms-march-2020

Vít, M., Olejník, R., Dlhý, J., Karpíšková, R., Cástková, J., Príkazský, V., Príkazská, M., Beneš, C. & Petráš, P. 2007. Outbreak of listeriosis in the Czech Republic, late 2006 – preliminary report. *Eurosurveillance*, 12: 3132. https://doi.org/10.2807/esw.12.06.03132-en

Whitworth, J. 2019a. U.S. sent meat from German firm linked to *Listeria* outbreak. Cited 12 October 2020. In: *Food Safety News*. Seattle Washington, USA. Cited 12 July 2022. www.foodsafetynews.com/2019/10/u-s-sent-meat-from-german-firm-linked-to-listeria-outbreak/

Whitworth, J. 2019b. Multiple meat items *Listeria* positive from 2017 in 2 country outbreak. In: *Food Safety News*. Seattle Washington, USA. Cited 12 July 2022. www.foodsafetynews.com/2019/11/multiple-meat-items-listeria-positive-from-2017-in-2-country-outbreak/

Whitworth, J. 2020a. Fatal *Listeria* outbreak linked to trout in the Netherlands. In: *Food Safety News*. Seattle, Washington, USA. Cited 12 July 2022. www.foodsafetynews.com/2020/07/fatal-listeria-outbreak-linked-to-trout-in-the-netherlands/

Whitworth, J. 2020b. Australia joins U.S. in having *Listeria* cases linked to imported mushrooms. In: *Food Safety News*. Seattle, Washington, USA. Cited 12 October 2020. www.foodsafetynews.com/2020/04/australia-joins-u-s-in-having-listeria-cases-linked-to-imported-mushrooms/

WHO. 2018. Listeriosis – Australia. In: *World Health Organization*. Geneva, WHO. Cited 12 July 2022. https://www.who.int/emergencies/disease-outbreak-news/item/09-april-2018-listeriosis-australia-en

WHO. 2019. WHO. 2019. Listeriosis – Spain. In: *World Health Organization*. Geneva, WHO. Cited 12 July 2022. https://www.who.int/emergencies/disease-outbreak-news/item/2019-DON256

Winter, C.H., Brockmann, S.O., Sonnentag, S.R., Schaupp, T., Prager, R., Hof, H., Becker, B., Stegmanns, T., Roloff, H.U., Vollrath, G., Kuhm, A.E., Mezger, B.B., Schmolz, G.K., Klittich, G.B., Pfaff, G. & Piechotowski, I. 2009. Prolonged hospital and community-based listeriosis outbreak caused by ready-to-eat scalded sausages. *Journal of Hospital Infection*, 73(2): 121–128. https://doi.org/10.1016/j.jhin.2009.06.011

Yde, M., Naranjo, M., Mattheus, W., Stragier, P., Pochet, B., Beulens, K., De Schrijver, K., den Branden, D., Laisnez, V., Flipse, W., Leclercq, A., Lecuit, M., Dierick, K. & Bertrand, S. 2012. Usefulness of the European Epidemic Intelligence Information System in the management of an outbreak of listeriosis, Belgium, 2011. *Eurosurveillance*, 17(38): 2–6.

Zhu, Q., Gooneratne, R. & Hussain, M. 2017. *Listeria monocytogenes* in fresh produce: outbreaks, prevalence and contamination levels. *Foods*, 6(3): 21. https://doi.org/10.3390/foods6030021

Current monitoring and surveillance programmes among different countries and regions

To determine unexpected vectors for food contamination, national CAs use a diversity of surveillance tools. These tools include establishing a microbial standard for *L. monocytogenes* for various food types, food product strategies, and surveillance strategies, along with analytical science and sampling tools.

The CA's *L. monocytogenes* surveillance system generally follows the implementation policies used for their national food safety system. Therefore, implementation can be based on the traditional inspection model with the CA taking full regulatory control, through to an industry-auditing system undertaken by third party auditors. It is not easy to find information on the specific national CA surveillance systems used to verify that the national *Listeria* food safety system is being effectively implemented.

The examples below are provided to give a perspective on the available monitoring programmes. The reader is reminded that some of the examples make reference to specific national and value chain contexts, and that the data presented here are not suitable to generalize the findings beyond the context of the example given here.

A3.1   AUSTRALIA AND NEW ZEALAND

Where applicable, food safety processors must implement an effective *L. monocytogenes* sampling plan for at-risk products, as well as implement environmental testing and be able to justify its sampling methodology for detecting *Listeria* spp. The business's food safety programme must include the frequency of testing, identify the size and location of environmental sampling sites, and detail corrective action procedures, including cleaning programme and handling of product following a positive test for *Listeria* spp. on a food contact surface (Australian Meat Regulators Group, 2016).

Challenge studies are required to provide information on the response of pathogens (growth and inactivation) to changes in the intrinsic physiochemical parameters and impact of production and processing factors. The design of a challenge study should adequately reflect the processes used to make the RTE food product.

Food regulators in Australia and New Zealand use a verification system to audit the food safety programme of licensed businesses. Research programme are undertaken after consultation with industry, with WGS used to explore *Listeria* problems and illness outbreaks.

A3.2 CANADA

It is the company's responsibility to ensure RTE products are safe and compliant, so they do require a preventive control programme for *L. monocytogenes*. The processing company's, premises and records are audited by Canada's CA.

Based on Codex principles, Canada's CA recommends that RTE industries implement environmental sampling programme with sample sites and numbers based on trend analyses. The CA provides guidance on establishing and monitoring control measures for *L. monocytogenes* in RTE foods. High risk products should be sampled at least monthly. Compositing of up to ten environmental samples is allowed, with *Listeria* spp. being the target organism to be analysed. It is recommended that tested food lots be held until all results are received. Clear guidance is provided to deal with any positive samples.

The Canadian Food Inspection Agency inspectors and third-party auditors verify industry compliance when undertaking sampling planned under the National Microbial Monitoring Programme and other targeted surveys. These samples are taken from retail and processing premises. Food products targeted include domestic cheese, imported dairy products, heat-acid rennet and coagulated cheese, egg products, RTE fruit and vegetables, pre-packaged salads, frozen fruit and RTE meat and fish, soy-based products, sauces and salad dressings, baked desserts, dips, nut butters, dairy flavoured milk and ice cream.

All regulatory tests must be undertaken at a government approved laboratory, with the *Listeria* methods used being those that appear in Health Canada's Compendium of Analytical Methods – Volumes 2 and 3 (https://www.canada.ca/en/health-canada/services/food-nutrition/research-programs-analytical-methods/analytical-methods/compendium-methods/laboratory-procedures-microbiological-analysis-foods-compendium-analytical-methods.html). Canada uses a range of science tools, including the Health Canada "Policy on *Listeria monocytogenes* in Ready-to-Eat Foods" enhanced listeriosis surveillance and WGS through the PulseNet Canada programme to manage food safety.

A3.3   CHINA

In China, the national government pays most of the costs of annual food safety surveillance, and every province must complete all of the work according to the national surveillance plan, and additionally, provide a real-time report of all detailed surveillance results to the central government (Pei *et al.*, 2015). Each year the government develops a national surveillance plan, issued in October before the implementation year. There are three types of microbiological surveillance: routine, special, and emergency surveillance. This surveillance plan selects the pathogen projects, which could include *L. monocytogenes*.

Food surveillance provides basic data and technical support for risk assessment and standards-setting, and promulgation of food safety laws. The surveillance results provide advice and tips for assessment projects, relevant analysis methods, and national standards of food safety. For example, the data regarding *L. monocytogenes* in RTE foods have been used in the food safety risk assessment project "quantitative risk assessment of *L. monocytogenes* in RTE food", where suggestions were provided to the classification and revision of "National Food Safety Criteria for Foodborne Pathogens".

In China, the first nation-wide surveillance on *L. monocytogenes* was conducted in 2000, supported by the Chinese Centers for Disease Control and Prevention (CCDC). Starting from 2010, the Chinese national monitoring network for microbial hazards in foods was set-up to survey all major foodborne pathogens including *L. monocytogenes* in 31 provincial regions (Pei *et al.*, 2015; Wu and Chen, 2018). Meanwhile, the national clinical listeriosis surveillance system in China was built in 2013 (Li *et al.*, 2018). Until 2020, there were no official reports or open-source datasets available that covered the national-wide findings on *L. monocytogenes* prevalence in different Chinese foods or clinical listeriosis. Thus, most of the data was provided by the partial findings of some provincial CCDC branches for periodic reviews (Li *et al.*, 2018; Liu *et al.*, 2020).

A3.4   EGYPT

In Egypt, food control functions are multisectoral; however, the main role in that area is carried out by the Ministry of Health and Population through its responsible bodies: (1) food safety and control department, (2) National Nutrition Institute (NNI), and (3) public health laboratories.

The enforcement of their national food regulations is administered by the food safety and control department at Ministry of Health and Population; such administration is necessary to ensure effective supervision and control and to take follow-up action as may be required. *L. monocytogenes* microbiological analysis is carried out by Central Public Health Laboratories.

A3.5   EUROPEAN UNION

In general, in the European Union (EU) and European Economic Area (EEA), there have been some improvements in the surveillance systems, in particular for countries with a relatively high level of reporting. However, there are still data gaps that make it difficult to make conclusions on the contributing factors that lead to cases of listeriosis. However, representative data has been collected across the European Union and European Economic Area using a harmonized sampling strategy suitable for surveillance over time on the (1) prevalence and concentration of *L. monocytogenes* in RTE foods, (2) consumption of RTE foods, (3) prevalence of underlying conditions in different risk groups by age and gender, (4) retail and home storage temperatures, and (5) *L. monocytogenes* virulence (EFSA, 2018).

In Europe, the European Centre for Disease Prevention and Control (ECDC) collects, analyses and disseminates surveillance data on 56 communicable diseases and related special health issues from all 27 European Union Member States and two of the three remaining European Economic Area countries (Iceland and Norway). In 2018, a study coordinated by the ECDC analysed 2 726 human *L. monocytogenes* isolates from 27 countries between 2010 and 2015. It found that slightly under 50 percent of the cases are sporadic whereas the remaining half of the cases cluster together. Around one third of the cases that were identified as part of a cluster affected more than one country, often lasting for several years. However, only two listeriosis outbreaks were reported in the European Union in 2016 and five in 2015, which suggests that many of them have gone undetected (De Waal, 2018).

On the other hand, the EFSA can undertake special studies to investigate any food safety pathogen. For example, in 2009, 2010–11 and 2014, the EFSA undertook sampling surveys focused on RTE products such as smoked and gravad (sugar-salt marinated) fish, heat-treated meat products, as well as soft and semi-soft cheeses. The purpose of these surveys was to ascertain the frequency of *L. monocytogenes* in retail RTE foods and to gain an understanding of general European Union compliance with the *L. monocytogenes* standard.

Within the European Union and the European Economic Area, there are also specific initiatives. Some examples are given.

In Austria, the Federal Ministry of Health (www.bmg.gv.at) has the overall responsibility for food safety and food safety legislation. It coordinates the activities of the food inspection authorities of the nine federal provinces and of the laboratories designated for analyses of official samples. Under the authority of the annual federal control plan, the authorities of the nine federal provinces carry out on-site inspections of enterprises and take samples. They are responsible for administrative measures and punitive actions in case of any violations of the law. Import controls for food of non-animal origin are carried out by food inspectors. Imported food of animal origin is controlled by border veterinarians.

In Denmark, the Danish Veterinary and Food Administration verifies *Listeria* surveillance and investigates processors associated with listeriosis outbreaks. WGS is used in outbreak situations.

One Health – Surveillance of Human Listeriosis – France (Ellis-Iversen *et al.*, 2019). In France, human listeriosis has been notifiable since 1999. Cases are reported to the French Public Health Agency (PHA), and human *L. monocytogenes* isolates are forwarded to the National Reference Center (NRC) for *Listeria* at the Institut Pasteur for genomic sequencing and core-genome Multilocus Sequence Typing (cgMLST) typing. Approximately 350 cases of human listeriosis are reported annually, of which 99–100 percent isolates are received at the NRC.

The NRC also receives food and environmental *L. monocytogenes* strains isolated from 1) "food alerts" from the Ministry of Agriculture (MoA) when food exceeds the regulation-defined threshold of *Lm*, 2) testing of food from the homes of patients with neurolisteriosis as part of the national surveillance strategy, and 3) food producers' own checks. All isolates from 1 and 2 are included in the National Surveillance System directly. Isolates from producers' internal checks are typed at the NRC based on private contracts and are only included if cgMLST matches a case. The surveillance system then has power to request disclosure of information from the producer including type of food, date of collection and name of the company to facilitate an investigation in a timely manner.

Additionally, the National Reference Laboratory (NRL) of the French Agency for Food, Environmental and Occupational Health and Safety (ANSES) collects isolates from food and environmental samples from national plans for surveillance and control of *Listeria* in the food chain and from other control programmes and surveys that are conducted annually to assess *L. monocytogenes* contamination of selected food items. These isolates are shared with the National Surveillance System once WGS-based typing has been implemented by NRL under the supervision of the European Reference Laboratory (EU RL) at ANSES.

The French national surveillance for listeriosis consists of epidemiological information on human cases from PHA, on microbiological surveillance of human, food and environmental samples performed by the NRC, and on samples from surveillance and control plans at the NRL. Both the Ministry of Health and the Ministry of Agriculture fund the surveillance activities. A "*Listeria* Unit" has managed the French national surveillance using a One Health (OH) approach since 1992 and is comprised of the Ministry of Agriculture, Ministry of Health, Ministry for the Economy and Finance, French Food Safety Agency (ANSES), NRC and NRL (www.fao.org/3/ab539f/ab539f.htm).

A cgMLST-based microbiological surveillance performed weekly by the NRC identifies clusters of matching isolates and shares these with the PHA and the MoA. Clusters involving at least one human case are jointly investigated by the PHA, the MoA and the NRC. Clusters that do not include human isolates are investigated by the MoA and the NRC. Merging of information from the PHA, the NRC and the MoA databases allows for efficient sharing of relevant data between agencies and timely investigations.

The close collaboration between the PHA, the MoA and the NRC has increased the number of solved clusters and outbreaks since 2015. In the future, the aim is to include food and environmental *L. monocytogenes* isolates received at the NRL in the National Surveillance System. The OH challenge is that each agency is responsible for their own databases on separate servers. Full integration would require a storage solution that allows for joint storage and sharing without compromising the integrity of the original data.

In Germany, representative data of *L. monocytogenes* in RTE foods is collected, assessed and published at regular intervals in the framework of the competent authority's zoonoses programme. The food products sampled include milk, soft, semi-soft and hard cheeses made from raw milk, tartar beef, spreadable/sliceable/cured meats, smoked and gravid fish, raw shrimps and fresh produce (sprouts, berries and vegetables).

In Ireland, all food business operators have a legal responsibility to produce safe food (Regulation 178/2002). The safety of foodstuffs is ensured by a preventative approach that is the implementation of a food safety management system based on the principles of Hazard Analysis and Critical Control Point (HACCP).

A3.6 LATIN AMERICA

The actual burden of listeriosis may be well underestimated in certain geographical regions due to lack of data. In this context, there is a gap in the estimation of the actual number of listeriosis cases in Latin America. This is most likely due to under-reporting because of a lack of specific surveillance and standard reporting of listeriosis in many Latin American countries. For example, while in countries like Chile and Uruguay the notification of listeriosis cases in humans is mandatory (although it is done through different pathways), in other countries such as the Plurinational State of Bolivia, Ecuador, or Peru, mandatory notification of listeriosis cases to the local health authorities or ministries is not required. In addition, several countries in Latin America have a passive surveillance of listeriosis through submission of *L. monocytogenes* isolates obtained from either foods or human cases, but not an actual reporting of the number of clinical cases. In many of these cases, even when information about the number of listeriosis cases per year, or statistics for passive surveillance is available, it may not be regularly updated or published by the local authorities of the country.

Reports of sporadic listeriosis cases in Latin America can be found in the literature, with many of them being reported in "grey literature" or journals that are not indexed.

A summary of the surveillance status of listeriosis for selected countries in Latin America is presented below:

TABLE A3. Surveillance status of human listeriosis in select countries of Latin-America

Argentina	• No mandatory notification of listeriosis cases • No mandatory notification of foodborne outbreaks • Passive surveillance of *L. monocytogenes* through the national reference laboratory (*Listeria* isolates from human cases and RTE foods)
Bolivia (Plurinational State of)	• No mandatory notification of listeriosis cases • No mandatory notification of foodborne outbreaks • Mandatory notification for gastrointestinal illnesses is for acute cases of diarrhea in children

(cont.)

Brazil	• No mandatory notification of listeriosis cases • Indirect surveillance through data obtained from mandatory notification of foodborne outbreaks (isolation of pathogens) • Reports of sporadic cases in the literature
Chile	• Mandatory notification of listeriosis cases since April 2020 • Mandatory notification of foodborne outbreaks (isolation of pathogens) • Passive surveillance of *L. monocytogenes* through the national reference laboratory (isolates from human cases and foods, submitted from other laboratories) • Reports exist for both outbreaks and sporadic cases in the literature • One report of the use of WGS
Colombia	• Only passive surveillance of *L. monocytogenes* through the national reference laboratory (from *L. monocytogenes* isolates submitted from the national laboratories network) • Reports of sporadic cases in the literature
Ecuador	• Mandatory notification of foodborne diseases; specifically, "hepatitis A", "salmonellosis", "typhoid and non-typhoid fever", "shigellosis", "cholera", and "Other Foodborne Poisonings" (the latter does not report the pathogen involved)
Peru	• Only mandatory notification of foodborne outbreaks • Reports of sporadic cases in the literature
Uruguay	• Surveillance of invasive listeriosis through mandatory notification of meningitis cases • A report from 2017 indicates an average of 3 cases per year, with an increase in the number of listeriosis cases (13 cases) in 2016; this increase in the number of listeriosis cases did not show an epidemiological link or common food sources. • There is no information available regarding the type of food involved in reported listeriosis cases.
Venezuela (Bolivarian Republic of)	• Only mandatory notification of foodborne outbreaks • Official epidemiological surveillance reports for communicable diseases has been suspended since 2017 (censorship of health data). • Reports of sporadic cases in the literature

The implementation of surveillance systems for listeriosis in humans is key to ascertain the actual burden of this illness and the role of specific food sources in LMICs. This knowledge can be a valuable contribution for the implementation of preventive measures throughout the food chain, as well as for the improvement in the risk communication of listeriosis for highly susceptible people and for the general population of these countries.

A3.7 SOUTH AFRICA

Monitoring programmes by food business operators

FBOs develop and implement risk-based environmental monitoring programme as part of their regulatory requirements on hygiene management system for *Listeria* spp. Each programme is specific to the plant and details the highest risk areas to be sampled for *L. monocytogenes* as determined by the FBO. The majority of FBOs were monitoring for *L. monocytogenes* and in some instances for *Listeria* spp. Environmental monitoring is used in conjunction with end product testing, not as a replacement for it.

The Foodstuffs, Cosmetics and Disinfectants Act, Number 54 of 1972 (DHSA, 2018) prohibited selling, manufacturing or importing any foodstuff which is contaminated or is in terms of any regulation deemed to be harmful or injurious to human health. No mandatory microbiological standards for foodstuffs were enforced on local products for *L. monocytogenes*. South African standard (Industry Standard) has microbiological requirements for *L. monocytogenes* limits applicable to processed RTE meat products at a maximum of 100 CFU/g during the shelf-life of the product. Recommendations for export of RTE products of animal origin has additional requirements as per import requirements by various markets for the absence of *L. monocytogenes* in 25 g for RTE foods destined for export.

Monitoring and surveillance programmes by competent authorities

A national monitoring and surveillance programme was implemented between 2014 to 2016 where 2 017 samples of raw and RTE food samples were collected and analysed for *L. monocytogenes*.

During the Listeriosis outbreak in South Africa

The 2017–2018 listeriosis outbreak in South Africa, caused by *L. monocytogenes* ST6 (CT4148), was traced to contaminated processed meats produced by Enterprise Foods, a subsidiary of Tiger Brands. Closure and a recall of all RTE processed meat products produced at the targeted facilities was instituted.

An amendment to the National Health Act (61/2003) (DHSA, 2017) was done so as to introduce compulsory notification of all listeriosis cases. The regulation defined "contamination" as the presence of an infectious or toxic agent or matter on a human or animal body surface, in or on a product prepared for consumption or on other inanimate objects, including conveyances, that may constitute a public health risk.

All districts in South Africa were requested to complete the profiling on all food manufacturing facilities. From this list and through site visits conducted, a complete list of facilities that produce RTE processed meat was obtained.

All production facilities that manufacture RTE processed meat in South Africa were identified (n=160), and all were inspected by district environmental health practitioners, supported by a core incident management team. A number of small production facilities with local distribution networks were identified as being contaminated with *L. monocytogenes* in the environment (not on food).

Inspectors conducted inspections of the targeted facilities and collected environmental swabs for the detection of *L. monocytogenes*. The number of samples were variable and based on the assessed risk at the facility. Reports and laboratory results were submitted to district municipalities for follow-up actions (if results were positive for *L. monocytogenes*) and for record-keeping in terms of ongoing inspection and monitoring.

At least 35 percent (n=160) of the facilities were identified as having non-ST6 (i.e. non-outbreak related) *L. monocytogenes* present in their post-heat treatment area.

Actions included serving of a prohibition notice, and/or compelling the manufacturers to conduct a "deep clean" of premises, and/or resampling, with ongoing intensive monitoring.

After the outbreak

South Africa tightened *L. monocytogenes* regulations for processed meat products by introducing a regulation under the National Regulator for Compulsory Specifications (NRCS). The regulation includes risk categorization of animal products including RTE and an amendment to the 2011 South African technical standard to be mandatory.

An absence (or 0 CFU/g) of *L. monocytogenes* in 25 g for RTE products that support the growth of *L. monocytogenes* tested in accordance with SANS 11290-1/ SANS 11290-2 at the end of manufacture or point of entry or point of sale during their shelf-life is proposed in the regulation.

L. monocytogenes is to be absent in 25 g for RTE products that do not support the growth of *L. monocytogenes* at the end of manufacture or the point of entry. At the point of sale during their shelf-life, the microbiological criteria will be < 100 CFU/g in 25 g when tested in accordance with SANS 112901/SANS 11290-2.

The frequency of *L. monocytogenes* testing is not defined; however, it is assumed that the samples must be representative and risk-based.

RTE food and environment surveillance

- Monitoring and surveillance of home/food establishments and food/environmental sampling is conducted by inspectors and environmental health practitioners.
- All *L. monocytogenes* isolates from food or environmental samples are to be kept by laboratories for routine confirmatory phenotypic testing and WGS.
- A routine national monitoring and surveillance programme was developed and is in the process of being implemented.
- There were no changes to the initial monitoring requirements of the FBOs.

A3.8 UNITED STATES OF AMERICA

The *Listeria* Initiative is an enhanced surveillance system that collects reports of laboratory-confirmed cases of human listeriosis in the United States of America. Demographic, clinical, laboratory, and epidemiologic data are collected using a standardized, extended questionnaire.

The *Listeria* Initiative was piloted in the Foodborne Diseases Active Surveillance Network (FoodNet) in 2004 and implemented nationwide in 2005. By 2014, the number of states participating increased to 47 and the District of Columbia. The proportion of all listeriosis cases reported to the *Listeria* Initiative continues to increase.

A main objective of the *Listeria* Initiative is to aid in the investigation of listeriosis clusters and outbreaks by decreasing the time from outbreak detection to public health intervention. Clinical, food, and environmental isolates of *L. monocytogenes* are subtyped. Since September 2013, state laboratories, the CDC, the FDA and FSIS have been performing WGS on all clinical, food, and environmental *L. monocytogenes* isolates. When clusters are identified, *Listeria* Initiative data are used to rapidly conduct epidemiological analyses. The food consumption histories of patients with cluster associated illnesses are compared with those of patients with sporadic illnesses to identify foods possibly associated with the cluster (CDC, 2016).

In the United States of America, WGS testing is used for the following purposes (FDA, 2017):

- to differentiate sources of contamination, even within the same outbreak;
- to determine which ingredient in a multi-ingredient food harbored the pathogen associated with an illness outbreak;
- to narrow the search for the source of a contaminated ingredient, even when the source is halfway around the world; and
- as a clue to the possible source of illnesses – even before a food has been associated with illnesses by traditional epidemiological methods.

A3.9 REFERENCES

Australian Meat Regulators Group. 2016. *Guidelines for the management of Listeria under Standard 4.2.3 the Production and Processing Standard for Meat.* St Leonards, NSW. www.pir.sa.gov.au/__data/assets/pdf_file/0006/238038/Guidelines_for_the_control_of_Listeria_6June2016.pdf

CDC. 2016. *National Enteric Disease Surveillance: The Listeria Initiative.* Atlanta, Georgia, USA. https://www.cdc.gov/listeria/pdf/ListeriaInitiativeOverview_508.pdf

De Waal, A. 2018. ECDC listeria surveillance: New EU-wide study reveals that most outbreaks remain undetected. In: *H5N1.* Cited 5 July 2021. https://crofsblogs.typepad.com/h5n1/listeria

DHSA (Department of Health of South Africa). 2017. *National health act. 2003 (act no. 61 of 2003). Regulations relationg to the surveillance and the control of notifiable medical conditions. Government notices No. 604 of 30 June 2017.* Pretoria, Republic of South Africa. www.gov.za/sites/default/files/gcis_document/201706/40945gon604.pdf

DHSA. 2018. *Foodstuffs, cosmetics and disinfectants act, 1972 (act no. 54 of 1972). Regulations relating to the hazard analysis and critical control point system (HACCP system): amendment. Government notices No. R607 of 14 July 2018.* Pretoria, Republic of South Africa. http://extwprlegs1.fao.org/docs/pdf/saf185428.pdf

EFSA. 2018. *Listeria monocytogenes* contamination of ready-to-eat foods and the risk for human health in the EU. EFSA BIOHAZ Panel (EFSA Panel on Biological Hazards). *EFSA Journal,* 16(1): 5134. https://doi.org/10.2903/j.efsa.2018.5134

Ellis-Iversen, J., Petersen, C. K. & Helwigh, B. 2019. *Inspiration and ideas - One health integration in surveillance.* Lyngby, Denmark, Technical University of Denmark. https://backend.orbit.dtu.dk/ws/files/183620811/Rapport_One_Health_Integration_in_Surveillance.pdf

FDA. 2017. Examples of how FDA has used whole genome sequencing of foodborne pathogens for regulatory purposes. In: *USFDA.* Cited 5 July 2021. www.fda.gov/food/whole-genome-sequencing-wgs-program/examples-how-fda-has-used-whole-genome-sequencing-foodborne-pathogens-regulatory-purposes

Li, W., Bai, L., Fu, P., Han, H., Liu, J. & Guo, Y. 2018. The epidemiology of *Listeria monocytogenes* in China. *Foodborne Pathogens and Disease*, 15(8): 459–466. https://doi.org/10.1089/fpd.2017.2409

Liu, Y., Sun, W., Sun, T., Gorris, L., Wang, X., Liu, B., & Dong, Q. 2020. The prevalence of *Listeria monocytogenes* in meat products in China: A systematic literature review and novel meta-analysis approach. *International Journal of Food Microbiology*, 312: 108358. https://doi.org/10.1016/j.ijfoodmicro.2019.108358

Pei, X., Li, N., Guo, Y., Liu, X., Yan, L., Li, Y., Yang, S., Hu, J., Zhu, J. & Yang, D. 2015. Microbiological food safety surveillance in China. *International Journal of Environmental Research and Public Health*, 12(9): 10662–10670. https://doi.org/10.3390/ijerph120910662

Wu, Y. N., & Chen, J. S. 2018. Food safety monitoring and surveillance in China: Past, present and future. *Food Control*, 90, 429-439. https://doi.org/10.1016/j.foodcont.2018.03.009

Annex 4

Reported food source attribution of listeriosis related to specific food groups

TABLE A4. Reported food source attribution of listeriosis related to specific food groups

	Type	Meat	Dairy	(Shell) fish	Fruits and vegetables	Other	Not food
FSIS/FDA	QMRA	90.9	8.8	0.2	0.06	/	/
Wambogo	Outbreaks	2.1	48.4	1.9	47.5	/	/
Batz (2014)	Outbreaks	45.0	30.0	5.0	5.0	15.0	/
Fillepello	Subtypes	35.8	44.4	6.2	/	13.5	/
Little	Subtypes	37.4	2.1	18.8	5.9	33.4	2.3
Havelaar	Expert	30.4	25.0	17.9	7.1	14.3	5.3
Davidson	Expert	58.2	26.7	6.1	8.4	0.5	/
Batz (2012)	Expert	59.9	23.6	7.2	8.7	0.7	/
Average	/	45.0	26.1	7.9	11.8	12.9	3.8

Sources: adapted from Batz *et al.*, 2012, 2014; Davidson *et al.*, 2011; FDA and FSIS, 2003; Filipello *et al.*, 2020; Havelaar et al., 2012; Little *et al.*, 2010; Wambogo , 2020.

REFERENCES

Batz, M.B., Hoffmann, S. & Morris, J.G., Jr. 2012. Ranking the disease burden of 14 pathogens in food sources in the United States using attribution data from outbreak investigations and expert elicitation. *Journal of Food Protection*, 75(7): 1278–1291. https://doi.org/10.4315/0362-028X.JFP-11-418

Batz, M., Hoffmann S., & Morris, J.G., Jr. 2014. Disease-outcome trees, EQ-5D scores, and estimated annual losses of quality-adjusted life years (QALYs) for 14 foodborne pathogens in the United States. *Foodborne Pathogens and Disease*, 11(5): 395–402. https://doi.org/10.1089/fpd.2013.1658

Davidson, V.J., Ravel, A., Nguyen, T.N., Fazil, A. & Ruzante, J.M. 2011. Food-specific attribution of selected gastrointestinal illnesses: estimates from a Canadian expert elicitation survey. *Foodborne Pathogens and Disease*, 8(9): 983–995. https://doi.org/10.1089/fpd.2010.0786

FDA & FSIS. 2003: Quantitative assessment of relative risk to public health from foodborne *Listeria monocytogenes* among selected categories of ready-to-eat foods. In: *USFDA*. Cited 20 June 2021. https://www.fda.gov/food/cfsan-risk-safety-assessments/quantitative-assessment-relative-risk-public-health-foodborne-listeria-monocytogenes-among-selected

Filipello, V., Mughini-Gras, L., Gallina, S., Vitale, N., Mannelli, A., Pontello, M., Decastelli, L., Allard, M.W., Brown, E.W. & Lomonaco, S. 2020. Attribution of *Listeria monocytogenes* human infections to food and animal sources in Northern Italy. *Food Microbiology*, 89: 103433. https://doi.org/10.1016/j.fm.2020.103433

Havelaar, A.H., Haagsma, J. A., Mangen, M.J., Kemmeren, J.M., Verhoef, L.P., Vijgen, S.M., Wilson, M., Friesema, I.H., Kortbeek, L.M., van Duynhoven, Y.T. & van Pelt, W. 2012. Disease burden of foodborne pathogens in the Netherlands, 2009. *International Journal of Food Microbiology*, 156(3): 231–238. https://doi.org/10.1016/j.ijfoodmicro.2012.03.029

Little, C.L., Pires, S.M., Gillespie, I.A., Grant, K. & Nichols, G.L. 2010. Attribution of human *Listeria monocytogenes* infections in England and Wales to ready-to-eat food sources placed on the market: adaptation of the Hald *Salmonella* source attribution model. *Foodborne Pathogens and Disease*, 7(7): 749–756. https://doi.org/10.1089/fpd.2009.0439

Wambogo, E.A., Vaudin, A.M., Moshfegh, A.J., Spungen, J.H., Doren, J. & Sahyoun, N. R. 2020. Toward a better understanding of listeriosis risk among older adults in the United States: characterizing dietary patterns and the sociodemographic and economic attributes of consumers with these patterns. *Journal of Food Protection*, 83(7): 1208–1217. https://doi.org/10.4315/JFP-19-617

FAO/WHO Microbiological Risk Assessment Series

1   Risk assessments of *Salmonella* in eggs and broiler chickens: Interpretative Summary, 2002

2   Risk assessments of *Salmonella* in eggs and broiler chickens, 2002

3   Hazard characterization for pathogens in food and water: Guidelines, 2003

4   Risk assessment of *Listeria monocytogenes* in ready-to-eat foods: Interpretative Summary, 2004

5   Risk assessment of *Listeria monocytogenes* in ready-to-eat foods: Technical Report, 2004

6   *Enterobacter sakazakii* and microorganisms in powdered infant formula: Meeting Report, 2004

7   Exposure assessment of microbiological hazards in food: Guidelines, 2008

8   Risk assessment of *Vibrio vulnificus* in raw oysters: Interpretative Summary and Technical Report, 2005

9   Risk assessment of choleragenic *Vibrio cholerae* 01 and 0139 in warm-water shrimp in international trade: Interpretative Summary and Technical Report, 2005

10 *Enterobacter sakazakii* and *Salmonella* in powdered infant formula: Meeting Report, 2006

11 Risk assessment of *Campylobacter* spp. in broiler chickens: Interpretative Summary, 2008

12 Risk assessment of *Campylobacter* spp. in broiler chickens: Technical Report, 2008

13 Viruses in food: Scientific Advice to Support Risk Management Activities: Meeting Report, 2008

14 Microbiological hazards in fresh leafy vegetables and herbs: Meeting Report, 2008

15 *Enterobacter sakazakii* (*Cronobacter* spp.) in powdered follow-up formula: Meeting Report, 2008

16 Risk assessment of *Vibrio parahaemolyticus* in seafood: Interpretative Summary and Technical Report, 2011

17 Risk characterization of microbiological hazards in food: Guidelines, 2009.

18 Enterohaemorrhagic *Escherichia coli* in raw beef and beef products: approaches for the provision of scientific advice: Meeting Report, 2010

19 *Salmonella* and *Campylobacter* in chicken meat: Meeting Report, 2009